Boundary Science

Boundary Science

Re-imagining Water-Energy-Food Interactions in the Context of a Data Light Approach to Monitoring the Environment-Development Nexus

Mathew Kurian

Consortium Lead-Belmont Forum
Project on Cyber-enabled Disaster Resilience
Pennsylvania
United States

Yu Kojima

Gender, Migration and Development Expert
Independent Consultant
Pennsylvania
United States

Elsevier
Radarweg 29, PO Box 211, 1000 AE Amsterdam, Netherlands
The Boulevard, Langford Lane, Kidlington, Oxford OX5 1GB, United Kingdom
50 Hampshire Street, 5th Floor, Cambridge, MA 02139, United States

Notices
Knowledge and best practice in this field are constantly changing. As new research and experience broaden our understanding, changes in research methods, professional practices, or medical treatment may become necessary.

Practitioners and researchers must always rely on their own experience and knowledge in evaluating and using any information, methods, compounds, or experiments described herein. In using such information or methods they should be mindful of their own safety and the safety of others, including parties for whom they have a professional responsibility.

To the fullest extent of the law, neither the Publisher nor the authors, contributors, or editors, assume any liability for any injury and/or damage to persons or property as a matter of products liability, negligence or otherwise, or from any use or operation of any methods, products, instructions, or ideas contained in the material herein.

Library of Congress Cataloging-in-Publication Data
A catalog record for this book is available from the Library of Congress

British Library Cataloguing-in-Publication Data
A catalogue record for this book is available from the British Library

ISBN: 978-0-323-88473-0

For information on all Elsevier publications visit our website at
https://www.elsevier.com/books-and-journals

Publisher: Candice Janco
Acquisitions Editor: Marisa LaFleur
Editorial Project Manager: Ruby Gammell
Production Project Manager: Kiruthika Govindaraju
Cover Designer: Victoria Pearson

Typeset by TNQ Technologies

Contents

Praise for Boundary Science

For a long time Nexus scholars have struggled to comprehend how governance influences Nexus discussions. In *Boundary Science* we discuss a two- step approach to integrating political economy analysis via the use of typologies of trade-offs and composite indices. This book uncovers the dire need to improve global public goods research through greater recognition of boundary spanning skills of political negotiation, research framing and horizon scanning in the environmental and social sciences. Boundary Science explores the applications of open access platforms to promote transdisciplinary engagement with pressing global challenges of migration and climate change and is an invitation to Nexus scholars and practitioners alike to reform the practice of environmental modelling by adopting open data principles to advance knowledge translation that supports robust decision making. The latest *World Development Report* 'Data for better lives' points out that management information systems within large bureaucracies are often ill-equipped to handle feedback on resource use and management of public infrastructure. This book crucially poses three questions that attempt to understand the political economy conditions that will lead to effective public action in support of environmental policy and management:

- By drawing upon the literature on incrementalism, we inquire about the pathways for integrating scientific inputs into environmental policy and planning (*the normative bases*).
- By relying upon an emerging body of work on transition planning and backcasting (in contrast to foresight analysis) we ask what approaches, frameworks and methods can support the piloting of policy instruments (guidelines, notifications, standards, circulars, directives) in environment and development planning (*the conceptual bases*).
- By reflecting upon agent-based models, we explore how the scope of data aggregation, collection and analysis/data processing modalities can be refined to provide more frequent, reliable, and disaggregated evidence that can advance the use of environmental models in decision-making (*the operational bases*).

In exploring the aforementioned questions, we coined the term *boundary science* to refer to an emergent body of literature encompassing theory, methods, models and data transformation techniques borne out of advancements in fields of public administration, political ecology, earth observations and data science. Boundary science is articulated via a discussion of three dimensions that have been overlooked by the literature on environmental policy and governance: (1) boundary conditions – we discuss seven of them in this book, (2) we delineate five typologies of trade-offs gleaned from global public goods research and (3) boundary spanning skills – horizon scanning, regional consultation, strategic communication, political negotiation and data valorization, which have hitherto been overlooked in the design and execution of environmental research.

The WEF nexus framework has the potential to be a powerful tool, yet it requires practical guidance for implementation. This book is a much-needed guide for nexus implementation that will not only advance scholarship but also facilitate on-the-ground use of nexus concepts in challenging, data-scarce contexts. Scholars, practitioners, and students will benefit from this book that guides them through the most challenging aspects of nexus thinking—bridging the gap between science and policy. Only by implementing nexus concepts and methods in real-world contexts can we discover the value of the approach and better understand how to advance nexus thinking for societal benefit.

Tamee Albrecht
PhD Candidate
University of Arizona

In a time of desperate global need, this book reconceptualizes the global disaster response framework by addressing water–energy–food (WEF) catalysts through a boundary science lens to understand the connections between environment and development. As developing countries are most affected by the repercussions of climate change, this book encompasses theoretical and practical pathways to provide an inclusive understanding of climate-induced displacement in a way that enables students and policymakers to reenvision the connections between environmental stress and socioeconomic and humanitarian issues.

Dr. Wa'ed Alshoubaki
The University of Jordan

For far too long, we, whether in university or government departments, have parsed the world into disciplines and sectors for our convenience when tackling a particular problem. However, the real world does not work in siloes. Each entity is a sum of its interactions with other entities. A nexus approach to research and policy enables us to rise above the siloes and appreciate the trade-offs, cobenefits, and feedbacks, which is essential for resolving the challenges of sustainability. This book is an important contribution to elucidating and applying a nexus approach to the water, energy and food systems.

Dr Maria Sharmina
Senior Lecturer in Energy and Sustainability
Tyndall Centre for Climate Change Research
School of Engineering
The University of Manchester

As a graduate student and social scientist in sustainability, this book provides a needed perspective on the intersection between sustainability science and the WEF nexus. A key challenge within the WEF nexus is integrating governance and nexus modeling approaches. The concept of boundary science presented here provides an opportunity to address this challenge by integrating governance and policy with the biophysical system through considerations of system scale and knowledge coproduction and through approaches to overcome challenges of missing data. The exploration of this integration in this book provides an opportunity to move towards the practical implementation of scientific WEF nexus data and research in policy.

J. Leah Jones
PhD Student, Arizona State University

Preface

The launching of the Sustainable Development Goals in 2015 was a breakthrough in connecting the 'global development agenda,' so prominent in the Millennium Development Goals (2000–2015), and the 'sustainability agenda', dealing with climate change, biodiversity loss and global environmental risk. Monodisciplinary studies, and its dominance in the institutional organization of many universities and of student exposure, had already given way to a more multi- and interdisciplinary approach in the organization of research endeavors dealing with major global issues. Many nonuniversity research and training institutes had already chosen a topical approach, which connected theories, models and data from a combination of disciplinary backgrounds, and some universities had also started a more issue-driven teaching and research setup, such as development studies, urban studies, information studies or environmental studies. And in some institutes, a nexus approach became en vogue: connecting a variety of issues, and the water–energy–food nexus has been a promising one.

However, the SDG, climate change, and biodiversity 'movements' also demand a transdisciplinary approach, where connections between 'knowing' and 'acting' and between science, policy, and what some people call 'citizen knowledge', or even 'citizen wisdom', become much more important. Boundary science is a concept exploring that interface. And it is not only about how scientific knowledge can be used, or better used in policy making and implementation. It is also the other way around: how behavior and behavioral change (by citizens, by politicians, by policymakers, and by NGO and business leaders) influences the way research is being conducted and knowledge communicated. Global social learning is important, as well as institutional resilience to political and societal resistance to major changes needed for the sustainability of the Earth's environment to support human and other life. Solutions demand cocreation (this book uses 'cocuration') of many different partners in and beyond the academic world.

Recent decades have shown the very rapid advances in Internet technology. Information sharing has become much more rapid and much more diverse. Smart phone technology makes information gathering much more easy, but also the use of 'citizen scientists' as cocollectors of data and co-designers of knowledge production. Combining different levels of scale, as well as combining 'big data' with a multitude of case study findings has become more easy, although grasping the organizational demands of it is still a major challenge. Online learning and online teaching had already become a rapidly developing global industry before 2020, but the COVID-19 pandemic has given it a major boost. Particularly for understanding the complex human–environment 'system' and its interconnectedness, the online learning that this book advances is an important step forward. And we can be sure: there is much more that will follow after this. And both the Earth's environment and humanity will need it ever more.

Ton Dietz
Emeritus Professor
Environment and Development Geography University of Amsterdam, and African Development Studies Leiden University

Acknowledgements

This book is a culmination of over two decades of engagement with water–energy–food (WEF) interactions spanning Africa, Asia, Europe and the Americas. By drawing upon case studies on participatory irrigation, wastewater reuse, monitoring of the UN Sustainable Development Goals, joint forest management and land use change, this book puts a spotlight on core governance challenges facing international development cooperation. Written in the midst of a global pandemic of historical proportions, this book questions why so many interventions aimed at improving the human condition have failed while other successes could not be replicated on a larger scale. This book in challenging key assumptions that have guided the WEF nexus debate such as those relating to population pressure, scale and data insufficiency calls for a broader focus on the environment–development nexus. The authors are grateful to the Belmont Forum for supporting the Theory of Change Observatory on Disaster Resilience (TOCO_DR) project that inspired this book (https://www.belmontforum.org/news/newest-belmont-forum-awardees-address-resilience-and-risk-in-disaster-scenarios/). This book has also inspired the launch of the Climate Panel (https://www.theclimatepanel.com/), an international initiative focused on training the next generation of leaders to understand and then act upon robust models of human–environment interaction. We are grateful to our partners in North America, Europe, Africa, South America, and Asia who have financed our research and from whom we have learnt so much by working together on several projects over the past two decades. This book also contains reflections of material that was prepared for undergraduate and graduate courses that were delivered to students drawn from different disciplines but with a common interest in issues of data, modelling and environmental policy and governance. We are grateful to the managers and editors at Elsevier for their patience and understanding during the process of preparing this manuscript for publication.

Mathew Kurian and Yu Kojima
Pennsylvania, United States

Prologue

In writing this book about the mechanics of sustainability science in the midst of a pandemic, we are forced to return to a question we raised several years ago in *Governing the Nexus*: Why does good science not result in good policy? The benefits of science become clearer now than ever before for its ability to predict, prepare and enable us to recover from risk for which society must rely not on epidemiological models alone but on its purported connections to public decision-making and communication. The pandemic has laid bare some of the underlying weaknesses of the scientific process: cognitive biases, slowness of the scientific method and the binary bounded nature of statistical modelling. Whereas the climate change debate on global versus downscaled models may seem abstract by contrast, the pandemic can help us to make sense of the disconnect between scientific modelling and public policy and political decision-making.

On April 5, 2020, the *New York Times* reported on how arm-chair epidemiology had succeeded briefly in offering explanations for a population that was seeking to come to terms with uncertainty brought upon by the pandemic. The Times article offered the example of an online post by a Silicon Valley product manager who argued against the severity of the virus, thereby suggesting that arm-chair epidemiology could offer a scientific-seeming explanation at a critical moment when scientists were contemplating the results of their trials. The explosion of social media and the 24 h news cycle has left us exposed to a stream of ever-evolving data that make it especially vulnerable to manipulation by hacks with adverse consequences for a critical resource we rely upon during a crisis – trust in science for its ability to predict, prepare and help us recover from a disaster. The store of public trust in the scientific process is depleted further whenever the scientific community is slow to respond or when decision-makers proceed based on conjecture and hyperbole. Either way when people get conflicting messages, it makes it hard for State and local authorities to generate the political will to take decisive action. The result can be catastrophic as can be judged by the number of deaths worldwide due to the pandemic!

This is a familiar picture for those engaged in sustainability science. To reverse this trend of eroding trust, it is time that we scrutinize closely the business of scientific process – how and why do we do environmental research? This book contends that public goods research has a responsibility to be policy relevant on account of the public nature of its financing. Here, we emphasize problem framing as a key point of departure – how do we view political dynamics involved at the levels of both cognition (how we/environmental scientists see the problem) and practice (how we/environmental scientists understand the function of the government and the modalities for advancing its interests)? Second, to address the issue of eroding trust, environmental research has a responsibility to contribute towards a paradigm shift in the design, appraisal, implementation, monitoring, and evaluation of development programs. Contributing towards incremental change in policy alone will not be sufficient to address some of the big environmental challenges of our time – for example, climate change. In similar fashion to a vaccine that was produced in record time, we need out-of-the-box thinking that will produce paradigmatic change in the way we approach the challenge of climate change to understand and then mitigate the critical trade-offs that arise from its effects upon other interconnected problems such as migration and forced displacement of people worldwide.

This means that we are in need of a renewed theory of change that can inform the design of integrative environmental models that draw upon longitudinal case studies to offer pathways on how the unintended effects of policy action may be mitigated. In response to this urgent call, we will draw upon our previous experience of designing, implementing and evaluating global public goods research to outline the contours of boundary science — an emergent body of theory, methods and tools of data transformation. This book will offer pathbreaking insights by identifying eight research design principles, discussing five typologies of environmental trade-offs and advocating for the large-scale use of environmental backcasting approaches via recourse to open access modelling platforms. As a result, we anticipate that *boundary science* will be useful for scholars of environmental science, development professionals and those with an interest in academic fields such as digital humanities, computer science, remote sensing, citizen science, economics, public finance, sociology and development studies.

Mathew Kurian, PhD
Yu Kojima, PhD
Pennsylvania
United States

CHAPTER 1

Why boundary science?

1. Introduction

The UN Sustainable Development Goals (SDGs) have generated immense interest among scholars interested in understanding the interdependencies among environmental resources and services and their governance. The examination of interdependencies has highlighted interactions between water, energy and food, especially as they relate to goals 2, 6 and 13. Two distinct patterns have emerged in research on water—energy—food (WEF) nexus interactions with implications for the SDGs in this regard. The first is focused on modelling of resource interactions with the objective of generating global scenarios based on current consumption trends. The second demonstrates the need for renewed emphasis on monitoring the quality, reliability and affordability of public services[1] such as irrigation or wastewater treatment. The latter emphasis on monitoring, underscores the need for an integrative framework that can accommodate for the need to downscale global environmental models and enhance their ability to better predict and prepare for risks associated with droughts and floods while at the same time being able to trace their effects on water, energy and food security (Kurian, 2020).

Against this background, this book offers a framework that can change the way we undertake development and scientific practice. From the vantage point of strengthening the institutional capacity[2] of the public agencies in relation to the use of evidence in development planning, there are three pressing priorities identified; 1) pathways for integrating scientific inputs into environmental policy and planning (normative basis for development planning), 2) integrative modelling that combines bio-physical and institutional perspectives (conceptual basis of environmental models) and 3) refining the scope for data and statistics for the purpose of

[1] Social services such as irrigation and waste water treatment could be provided entirely by the private sector. However, it is important to stress that the scope of this book is on delivery of services by the public sector in partnership with private contractors.

[2] Hill (2006) emphasizes the significance of capacity when she points to the assertion of Amartya Sen that equity in the exercise of institutionalized power involves more than increasing the input of individuals into social decision making; it involves the empowerment of individuals through their social organization and through increasing their self-determination in all areas of activity.

Boundary Science. https://doi.org/10.1016/B978-0-323-88473-0.00001-4

more effective monitoring of development outcomes and impacts (operational procedures and evaluation metrics). It is in this context, we introduce the new concept of Boundary Science which is premised on the environment-development Nexus. We locate our discussion on environmental governance in the context of the environment-development nexus by shifting the focus beyond poverty reduction debates and attempt to include analysis of broader institutional concerns. This implies that our analysis throughout this book will shed light on issues not limited to efficiency but extend to addressing those of equity such as the interventionist and partisan assumptions that neglect differentiation in distribution of benefits and political judgements that reinforce asymmetries in bargaining power as is often apparent in environmental governance.

In this vein, Boundary science, engineered by citizen science, aims to contribute to enhancing the effectiveness of development practice and rigor of the scientific enterprise by encompassing: (1) the intersections between planetary- and administrative-scale perspectives on environmental planning and management and (2) Stronger and clearer connection between theory, models and data to advance institutional analysis with the prospect of enhancing political buy-in for the outputs of public goods research (we call this in this book: the boundary spanning skills of horizon scanning and research framing). Along this line, we will demonstrate that the architecture of Boundary Science animates through the book via three propositions; (a) a theoretical proposition illustrates how the concept of WEF Nexus as a product of a multi-disciplinary approach stems from emerging critique of uni-dimensional models of water-energy-food (WEF) interactions combined with debates surrounding common property resources, agent-based modeling/agency behaviour analysis and social network analysis, (b) methodological proposition illuminates how the trade-off analysis which offers vivid features of divergences between environmental norms and institutional practice is underpinned by the use of composite indices and typology analysis and (c) operational proposition elaborates how to employ fit for purpose open-access modeling platforms that harness the applications of data transformation and visualization tools for adept policy decision-making.

One key point we emphasize throughout this book is the omni-presence of political economy in the process of public policy planning. This is evident for decades in the case of developmental planning interventions where often dynamic courses of interactions between institutional and environmental outcomes are constantly shaped by diverse voices and responses of public service users along the axis of gender, class and ethnicity in a rather erratic way. This stands in stark contrast to conventional operational norms and modalities that are yet to address an overly instrumentalist view which confines development planning within a linear and monotone straitjacket. This point brings us to the realization that the prospects of development planning could be enhanced if instead, such interventions are designed to embrace divergent pathways of development and their consequences. The lessons that we glean from such interventions can offer context specific insights and rich empirical experiences that can serve as a basis for building greater institutional resilience by capturing recursive effects through timely feedback and more effective coordination via responsive management information systems to overcome siloes in development planning. Given this scope, boundary science as a new analytical perspective is not limited to environmental science. Due to its conceptual fluidity and versatility, Boundary Science promises to offer new insights on methodology and tools for monitoring and evaluation that may be of relevance to other thematic areas of global public goods research. To this end, an expanded concept of environment and development Nexus will be examined from the perspective of the changing landscapes of environmental governance where features of current environmental policy challenges (e.g. climate change) increasingly intersect with issues surrounding equity in the context of development, i.e., gender, intersectionality, hierarchy in knowledge production and citizenship that are however, conventionally excluded from the focus of environmental research and regulatory initiatives related to environmental governance. We thus foresee that the promotion of an inclusive analytical framework such as the one promoted in this book has the potential to enhance cooperation among multiple public agencies and scientists drawn from diverse disciplinary backgrounds.

We outline in this chapter some of the pathways towards achieving paradigmatic change in the study of human—environment interaction. A central tenet of our thesis on paradigm change is predicated upon the view that change in the human-environment system should be planned for in phases marked by transition points that can be mapped out in advance via the use of back casting approaches. This would stand in stark contrast to foresight approaches that view change in operational terms of scenarios, performance benchmarking and experimental trials of technical and management options. In sum, some of the key chapter objectives are as follows:

- To understand the role of trade-off analysis in supporting re-conceptualization of water, energy, and food Nexus framework
- To outline a rationale for why public goods research should engage in social learning and knowledge translation through the provision of open-access platforms for modelling and online learning (in contrast to a focus on case studies and global/regional models)
- To introduce concepts of nonlinearity, nonmonotone, and recursive effects of change processes and their role in creating and sustaining blind spots in environmental governance

2. Re-imagining Water-Energy-Food (WEF) interactions in the context of a growing appreciation of coupled human-environment systems

Since the early 1970's, there has been a growing interest in environmental sustainability. Environmental nongovernmental organizations (NGOs) such as Green Peace have mobilized activists to protest whale hunting and raise public awareness about issues such as global warming and rights of indigenous people in threatened habitats such as the amazon forests. Influential environmental commissions compiled the Club of Rome and Brundtland Report to document the effects of rapid environmental decline and chart a more sustainable development pathway.[3] More recently, the United Nations convened the 1992 summit on Sustainable Development and in 2017 convened the Paris Summit on Climate Change. The Sendai framework and Hyogo protocols and the newly established Green Climate Fund (GCF) headquartered in Seoul all aimed to ensure that Member States committed to specific steps for ensuring environmental sustainability. Alongside global diplomacy, there have been several initiatives that have sought to provide the scientific basis for sustainable development—the International Panel on Climate Change (IPCC), the International Forestry Resources and Institutions (IFRI) project at Indiana University and the Global City Indicators Facility housed at University of Toronto to name a few examples (see Fig. 1.1).

This book itself is meant to inform curriculum that will be developed as part of a project on cyber-enabled disaster resilience that will be supported by the Belmont Forum.[4] The Belmont Forum, which is an influential consortium of international science foundations and donors, aims to promote sustainability science. Sustainability science encompasses work that aims to promote cooperation across disciplinary

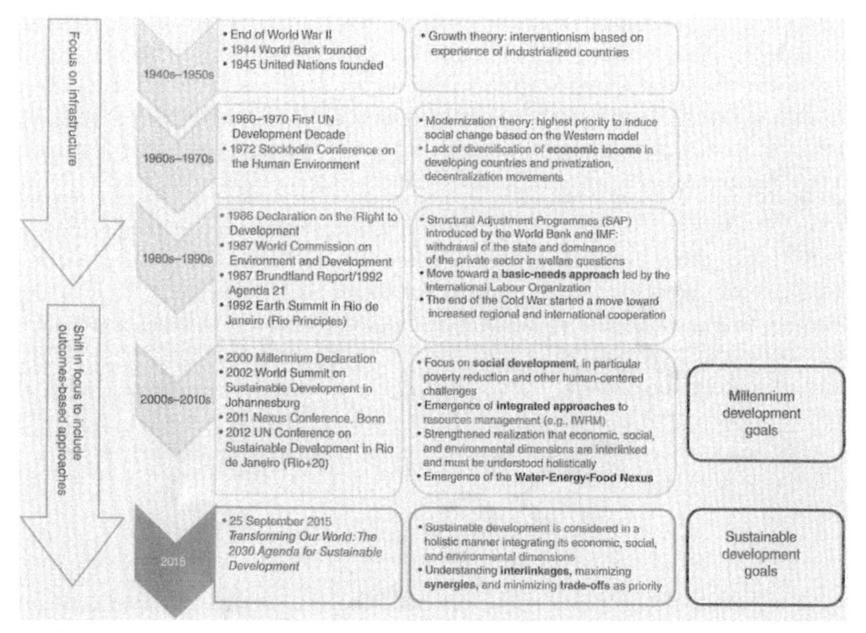

FIGURE 1.1

Chronology of international development trends. Source: Meyer and Kurian, (2017).

[3] For reference to the Bonn Conference of 2011, see Giz and ICLEI (2014).

[4] For more information on the Theory of Change Observatory on Disaster Resilience (TOCO_DR) project, please see https://www.belmontforum.org/news/newest-belmont-forum-awardees-address-resilience-and-risk-in-disaster-scenarios/.

boundaries and by engaging experts and nonexperts in finding solutions to global environmental challenges such as droughts, floods, and food insecurity. In 2019, the Belmont Forum formulated the relationship between humans and the physical environment as "**Extreme** environmental events that negatively impact **coupled** human-natural systems, including but not limited to impacts on economic, health, infrastructure and social subsystems. Extreme environmental events may be generated by **natural forces**, including climate change and or **anthropogenic causes**." For purposes of outlining the scope of this book, there are several aspects of the Belmont Forum formulation that are worth emphasizing. These include the notion of extreme events, coupled systems, natural forces, anthropogenic causes, and infrastructure; the examination of each could serve to further clarify the broad contours of this book (see https://www.belmontforum.org/).

Extreme environmental events (some of them induced by climate change) could give rise to risks such as landslides arising from high volume of rainfall in short periods, flooding arising from peak river discharges, severe storms often near coasts, heat waves, risks related to droughts—dry spells in "normal" wet seasons and frost—particularly when unexpected such as in early autumn or late spring seasons (Dietz, 2010). The frequency, duration, and intensity of extreme events could exacerbate the interdependence between natural forces and anthropogenic causes. For example, rising frequency of floods and peak river discharges could be the result of significant changes in land use that have occurred in the area as is the case in many parts of sub-Saharan Africa (Twisa et al., 2020). All these aspects may have negative impacts on water management and calls for adaptation measures to cope with lower predictability and more extremes. Adaptation means a shift to more water sources, from a wider geographical environment (more interdependence). It also calls for better defense mechanisms against extreme events, both technically and institutionally (Kurian et al., 2013). It is in this connection that a range of infrastructures—storm drains, irrigation canals, and wastewater treatment plants—become critical. Here the effect of three conditions in mediating the relationship between extreme events and adaptation via infrastructure needs elaboration: (A) bounded versus unbounded systems, (B) formal versus informal systems and (C) data rich versus data poor institutional environments (Kurian et al., 2019).

Bounded systems make it possible to procure resources such as water, energy or food to meet demands for public services locally. By contrast, systems for which resources such as water, energy or food have to be procured from outside a predefined physical or administrative boundary (i.e., river basin or municipality) to meet demands for public services may be referred to as unbounded systems (Gregersen et al., 1989). It is important to note that historically with the emergence of modern states, the credibility of authority and governance structures was sought to be enhanced through formal regulations that determined how central transfers (from central to provincial governments), taxes and tariffs could be used to operate and maintain infrastructure that provided critical services such as wastewater treatment, water supply and irrigation. As a result of growing formalization in management systems for provision of public services and infrastructure, seasonal risks related to changes in the weather were significantly reduced. However, with growing formalization of authority structures and management systems, cascading institutional risks are bound to be exacerbated due to growing interconnections between resources and multiple use services. The coupling of biophysical and institutional systems such as energy systems in densely populated cities, for instance, necessitates integrative modelling so as to adequately capture the effects of tariff regimes, seasonality and locality on patterns of water use (see Table 1.1).

With increasing coupling of biophysical and institutional systems, finding solutions to technical challenges should not be the sole focus of executive action. Instead, managing critical trade-offs is crucial to maintaining environmental resilience while at the same time ensuring that the needs of a populace for critical public services via appropriate choice of infrastructures are achieved. Nevertheless, predominant thinking on WEF interactions tends to promote a sectoral view, and thus it does not entertain examination of trade-offs between divergent interests and resources (budgets, skill set and staff of different public agencies) in the context of environmental planning which ultimately results in unmitigated trade-offs between environmental and social outcomes. In other words, a non-monotone, dynamic and robust view of the WEF Nexus instead would illuminate the various dimensions of political economy of environmental governance by securing the environment which ensures mitigation of trade-offs through overcoming of siloes and, sustaining necessary thresholds to public action via critical mass of financing and technology. We refer to this state as "synergies" (see Fig. 1.2).

The key concepts of siloes, thresholds and critical mass have mutually reinforcing effects to this end. First, in simple terms, siloes refers to an institutional state where obstacles exist to sharing information that prevent cooperation between public agencies and their staff. Secondly, thresholds to public action[5] is a state where the public agencies are effectively committed to an institutional environment that

Table 1.1 Example of Los Angeles Tariffs.

Type of family	Price in low block (USD/CCF)	Switch-point	Price in high block (USD/CCF)	
			W	**S**
Residential single family	$ 1.14	575 gallons/day	$ 2.23	
		725 gallons/per day		$ 2.98
Multi-family	$ 1.14	125% of winter U	NA	$ 2.92
Non- residential	$ 1.21	125% of winter U	NA	$ 2.98

Hannemann, M., 2000. Price and rate structures. In: Baumann, D.D., Boland, J.J., Hannemann, W.M. (Eds.), Urban water demand management and planning. Mc Graw- Hill, New York. pp. 137–179.

[5] Picketty (2020) argues that public investment serves to legitimize rules regarding the boundaries of a community and its territory, the mechanisms of collective decision making and the specific rights of its members (citizens) in contrast to its non-members.

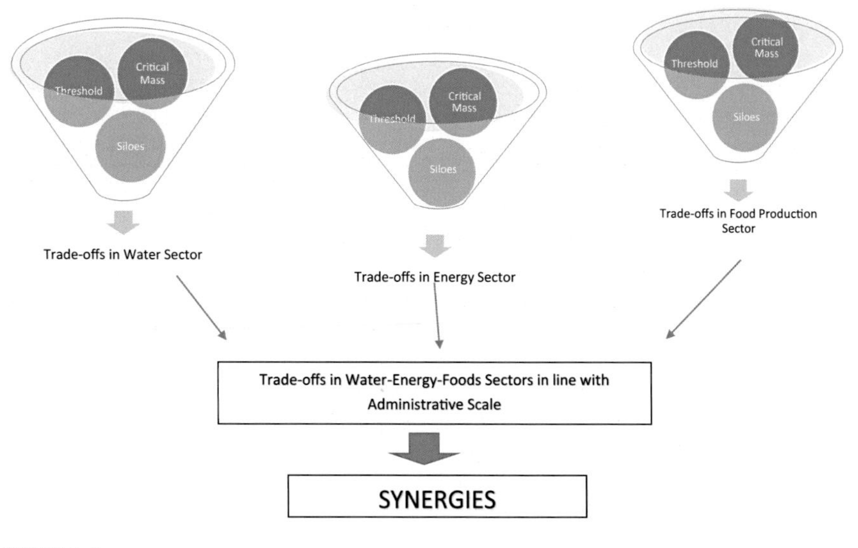

FIGURE 1.2

Three key components of synergies of WEF Nexus.

Source: Kurian and Ardakanian (2015).

promotes adequate staffing, skill development and policy support for coordination in decision-making. Third, critical mass connotes the state when the appropriate technology can be financed using available resources.

Let us examine these concepts further with an example from the transport sector. An assessment of three transport options (immersed tunnel, a bridge and semi underground tunnel) in the Netherlands revealed that managing trade-offs between congestion and delays for commuters were critically influenced by the costs of construction and maintenance of road projects (Reddy and Kurian, 2015). Therefore, greater attention to opportunities to mitigate trade-offs through examination of WEF Nexus considerations relating to siloes, thresholds and critical mass could have unearthed options for cost reduction and the advancement of public welfare. In this connection historical institutionalist literature enables us to identify three components of robust synergies that are key to appropriate choice of infrastructure for managing environmental trade-offs: (1) social networks that support information flows and knowledge exchange among different functionaries within and across departments, ministries and agencies, (2) deployment of complimentary skill sets (capacity) by key players and (3) a critical mass of financing and technology that can be appropriated by agencies and departments focused on achieving a particular policy goal (Gregory, 1997; Batley, 2004) (Box 1.1). For example, Box 1.1 illustrates how these three components interact to influencing infrastructure financing for delivery of urban sanitation services in Indonesia.

There are also several enabling factors for robust synergies, notably (1) a clearly articulated legal and policy framework, (2) clear set of policy instruments for implementation of legal and policy framework that includes directives, guidelines,

Box 1.1 Funding sources for urban sanitation, indonesia.

Sources of Funds:

Sanitation services can be developed with various sources of finance. Besides its own municipal budget, the city can access government and nongovernment funds. To reach the 2015 targets of the Millennium Development Goals, Indonesian cities need to access an additional IDR 4 billion annually from sources other than municipal budgets.

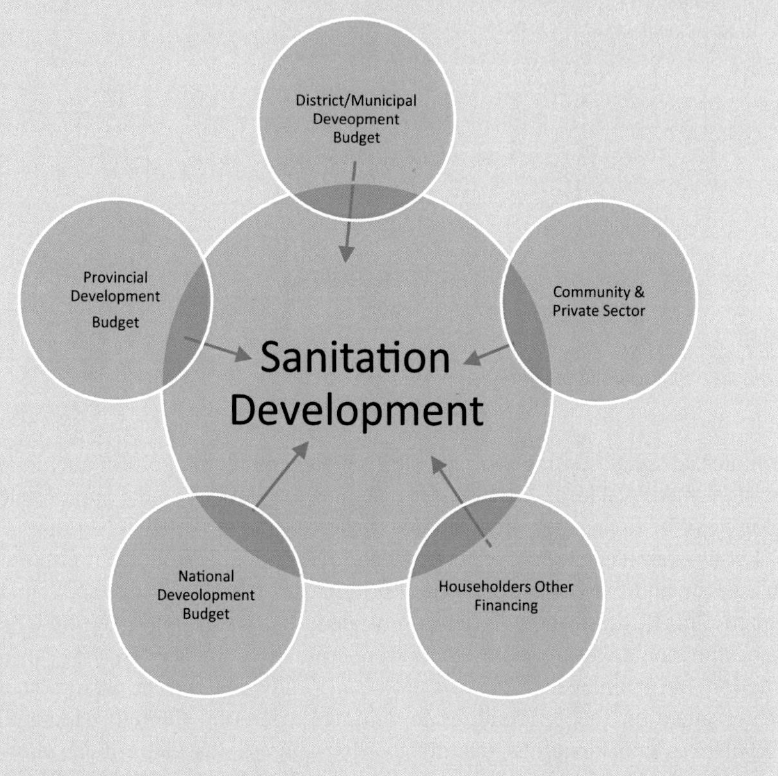

Source: *Adapted from WSP. (2009). Urban Sanitation in Indonesia: Planning for Progress. Jakarta, Water and Sanitation Program: 7.*

circulars, standards and notifications, stipulating how choices regarding technology and financing options may be arrived at, (3) data and evidence on distribution of biophysical and institutional risks, (4) manageable levels of administrative discretion with regards to interpreting and implementing policy instruments and (5) incentive structure (penalties and rewards) for compliance with policy instruments (Pollitt and Bouckeart, 2000; WSP, 2009; Kurian et al., 2018; Kurian and Ardakanian, 2015).

Here, it is important to appreciate the importance of feedbacks between environmental hazards (frequency, duration and intensity) and decision-making that focusses on managing institutional trade-offs. This is because it can demonstrate how executive decisions are influenced by interventions in the field. It is in this connection that data about actual and future state of the environment are critical to maintaining a robust link to decision-making (Mannschatz et al., 2015). Data fusion, integration, visualization and valorization are all key concepts to understanding the role of feedback loops in environmental decision-making. The role of big data, remote sensing and high-resolution computing must be emphasized in this context (see Fig. 1.2). Therefore, provision of adequate data infrastructure to enable data collection using a variety of medium and sources, data analysis combining varied disciplinary expertise and data transformation techniques that enabled co-curation and co-design of research is key to managing trade-offs in environmental decision-making. Indeed, adequate attention to the issue of data infrastructure will provide a context to research and enhance its ability to engage with real-world environmental challenges. But it is crucial to realize that data and information alone will not be sufficient—a complimentary understanding of individual and agency behaviour will also be necessary to successfully mitigate environmental trade-offs.

3. Why behaviour matters in shaping environmental change processes?

We pointed out in the previous section that with increasing coupling of biophysical and institutional systems, finding solutions to technical challenges are no longer the focus of executive action. Instead, managing critical trade-offs is crucial to maintaining environmental resilience while at the same time ensuring that the needs of a populace for critical public services via appropriate choice of infrastructures are achieved. Given the interdependence and associated complexity of socio-ecological systems, it becomes imperative that we pay careful attention to what aspects of an environmental project or program we choose to monitor to understand its potential to impact upon institutional outcomes. It is in this context that it becomes imperative to discuss how integrative environmental models can become a tool for monitoring public policy outcomes. In this connection, there are three areas which blind spots in environmental goverance are created and sustained; 1) (standardized/diverse) patterns of responses of consumers/users and institutional/environmental outcomes i.e., adoption rates of randomized controlled trials (RCTs) technical and management options - (non)monotone; 2) (fixed /erratic) course on the interaction between environmental resources and institutions- (non)linear and 3) (well/ill-coordinated) impacts of institutions/agency behaviour (reflected in budgetary, strategies, staffing and information sharing) on environmental outcomes- (non) recursive[6]

In contrast to perspectives offered by bio-physical scientists, the examination of the above three aspects of environmental models should help us appreciate the fact that in reality, environmental change processes are non-linear, non-monotone and recursive. The issue of non-linearity is borne out by the example of groundwater

[6] For more detailed description related to the issue of (non)monotone-(non)linear and (non)recursive discussion, please see Chapter 3: typology of trade-off analysis.

management. Feedback about the interaction of environmental resources usually assumes that where water is impounded; for example, in a tank, the chances of it percolating to increase recharge of groundwater may be enhanced. But our previous research in South India showed that groundwater recharge is dependent on the condition of the tank bed. Siltation can seal the tank bed and limit infiltration (Turral and Kurian, 2010). Similarly, we found that afforestation programs usually assume that downstream water users in urban areas would benefit from improved water supplies. Typically, in semiarid conditions, values for annual runoff vary from 5% to 15% of annual average rainfall, with greater amounts after heavy rainstorms on wet (saturated) soils and with relatively higher rates at the plot, field and microwatershed scale as compared with the sub-basin or basin scales (see Fig. 1.3). Here, our research revealed that replanting upper catchments with trees tends to reduce runoff by retaining more water and allowing less to flow downstream. In sum, implications of non-linearity or erratic course in the context of environmental goverance here is that the conventional assumption regarding the link between afforestation and urban water supply being linear is often challenged by unforeseen uncertainties contained in biophysical processes.

The issue of non-monotone effects between individual behaviour of consumers and environmental outcomes is borne out by experiments/field trials on technology options and management strategies. We found through previous research in the Mekong river basin that interventions that aimed to reverse soil erosion and capture

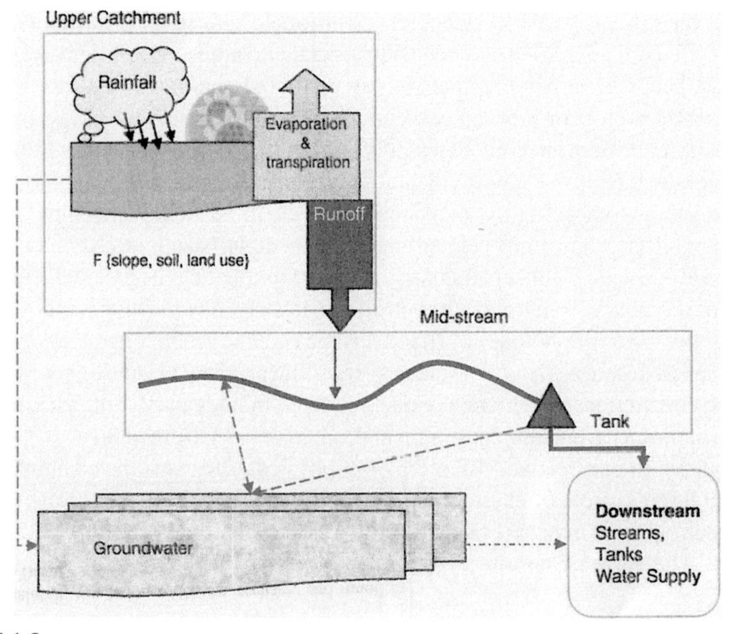

FIGURE 1.3

Schematic relationship between rainfall, recharge and run off in small catchment.

Source: Turral and Kurian, (2010):180.

soil nutrients often assumed that slash and burn agriculture had to be replaced by technologies such as improved fallow (IF). But agronomic assessments that were employed to convince farmers to adopt more sustainable land management practices had to contend with results that showed that although IF reduced soil erosion rates, they lowered crop yields by as much as 500 kilos per hectare depending crucially on where the plots were located (upper, middle or lower regions) and the extent to which they were already eroded (Kurian, 2010). Later in this book, we will discuss at some length another example that showed that contrary to conventional wisdom at the time, private entrepreneur led collective management strategies produced better social and environmental outcomes during the start-up phase of a participatory watershed management program compared with a cooperative-led management strategy. Here, these case studies demonstrate diverse outcomes of technical (first example) and information management (latter example) options. In this context, non-monotone implies unintended outcomes of the developement intervention reinforced by the lack of robust feedback loops between a development intervention and public policy outcomes. This could explain the reason for the poor adoption of technology adoptions recommended by global public goods research because of their narrow focus on only enhancing agricultural production without referring to its implications on market and trade policy in the case of Laos (Renkow, 2018).

The recursive effect of institutions on environmental outcomes is an issue that is complex and has received relatively limited attention by global public goods research. Elinor Ostrom's (1990) critical distinction between constitutional rules (example, property rights), collective action rules (example, budget appropriation rules) and operational rules (example, policy directives and guidelines) can help us appreciate the complex effects of institutions on environmental outcomes (see

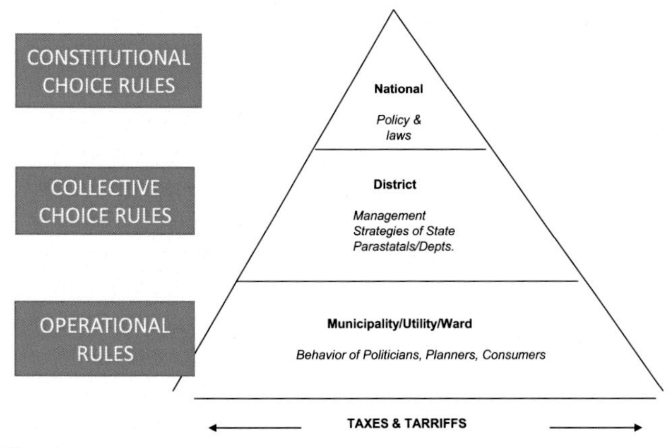

FIGURE 1.4

Institutional environment.

Fig. 1.4). Ostrom points out that because constitutional choice rules are more difficult to change, environmental management interventions that necessitate changes at the level of constitutional choice (for example, changes in land zoning and land use classification) would benefit from continuous updating of management information systems to reflect the changes in the wider political economy. For example, in Chapter 3 we rely on an example of participatory watershed management to show that when markets for forest products shrunk due to changes in India's export policy, farmers were no longer interested in investing in forest protection because it had become less profitable. Here again, non-recursive effects of institutions on environmental outcomes connotes the significance of coordinating interventions across multiple sectors in view of changes in the larger political and economic landscape.

To conclude, it would not be wrong to assert here that institutional resilience is advanced when governance systems have developed the capacity to address the effects of non-linearity, non-monotony and recursiveness in the design, implementation, monitoring and evaluation of public policy interventions. In other words, as long as the government organizations are able to maintain a well-coordinated and regularly updated institutional feedback mechanism (recursive), institutional and environmental outcomes arise from diverse response of consumers/users (non-monotone) and dynamic course on the interaction between environmental resources and institutions (non-linear) could actually be considered as the great asset for building institutional resilience to address environmental challenges. This is the reason why we propose open-access platforms as an effective operational tool to strengthen the institutional coordination capacity through more effective and responsive information management. In this connection, the example of the public documentation innovation and service (PDIS) initiative in Taiwan (https://wwwpdis.nat.gov.tw/en/) emphasizes the importance of moving to up to date/real time sources of information that can guide mobilization in support of public action. The implication of digital information initiatives such as PDIS is that reform of less periodic information and associated institutional practices of country census and statistics bureaus has become imperative. This would imply that we would need to develop a new paradigm of environmental change research with the theoretical and political considerations which we highlight in Table 1.2 below.

Bjorkman (2010) and Veiga et al. (2015), emphasize the ubiquitous nature of the budgetary process as being responsible for subverting the larger scale impact of environmental interventions. "The standard practice is to proceed from a historical base, to accept the recent past as a given to concentrate on proposed new increments to the budget. Because of this complexity, the actors use mechanisms or devices such as padding, across-the board cuts, accelerated spending at the end of the fiscal year, and often a refusal to release a proportion of budgeted funds until the last possible moment" (Bjorkman, 2010:167). Such practices can result in a tendency of fragmented budgeting where "special funds" are taken "off budget" or outside the annual cycle of budgeting with the objective of protecting them from political manipulation. However, the plethora of autonomous funds means that the budget loses control of revenue and expenditure and can no longer guide the behaviour of government.

Table 1.2 New Paradigm of Environment Change Research Incorporating Political Economy Perspectives.

Features	Threshold	Critical Mass	Siloes
Function	Institutional Capacity	Financing and Technology	Data sharing
Related Theory	Agent-Based Modelling (ABM) Agency Behaviour Analysis (ABA)	Common Property Resources (CPR) Institutional Analysis and Development Framework	Water-energy-Food Interactions Social Network Analysis (SNA)
Monitoring	Policy & Legal Framework Policy Instruments (Directives, Guidelines, Notifications, Standards, Circulars)	Finance & Budgeting	Design (Research and Programme)
Non-Linear	Incentive and budget structure not aligned	Budget constraints	Project/ budget cycle not aligned
Non-monotone	Skills and roles not synchronized	Access to budget with no/ wrong technology choice; no access to budget and technology available	Multi-faceted information not collected
Non-Recursive	Unintended effects of policy not mitigated	Feedback loop between expenditure and outcomes are not robust	Interventions are not responsive to feedback

Another challenge that the budget process can pose to monitoring environmental outcomes that are especially pertinent to the global south is that successive governments can commit resources towards capital expenditures to capture short-term electoral gains but without setting aside sufficient resources for recurring expenses such as salaries and operation and maintenance costs of established infrastructure that have the potential of advancing sustainable development.

4. Learning to learn: The framing challenge in sustainability research

This book promotes the idea of boundary science. Whereas the theoretical and operational concept of boundary science will be discussed at length throughout this book, in simple terms, we define boundary science as the new way of advancing integrative analysis of scale and boundary conditions in environmental research to pursue sustainable development. One of the benefits of adopting Boundary science is that it allows us to capture clearly what we are up against - the gulf that exists between two dominant standpoints of sustainability science and the WEF nexus. Whereas sustainability science, as adopted promotes cross-disciplinary and multi-sectoral collaboration as a way to enhance the relevance, use and impacts of research, the WEF debate on the other hand focusses on understanding the interdependencies emerging from their management and use of environmental resources and potential to mitigate trade-offs through synergestic action (Belcher et al., 2019; Kurian et al., 2019). Integrative models that promote the use of interoperable and anonymized data sets, expert panels and pilot-testing of policy instruments can go a long way in bridging the gap that exists between the WEF Nexus and sustainability science.

The framing challenge is mirrored in the siloed perspective that governs how we engage in environmental research. Siloes in environmental research are reinforced by disciplinary differences in language and interpretation of results, limited exposure to undertaking multidisciplinary research and low value attached to skills of partnership building and political negotiation in policy-oriented research. For example, suppose we choose the topic of water quality, in the biophysical domain, questions may be focused on understanding how water quality affects human health.[7] On the other hand, in the social science domain, research may focus on understanding how community collective action may help improve water quality by lowering the cost of monitoring. Boundary science is premised on the idea of bringing these two binary sets of research/disciplinary perspectives to frame a question that is focused on understanding the institutional trajectory of policy and management interventions that advance the goal of achieving water quality.

In practice, boundary science would necessitate the pursuit of biophysical and social science research via the use of a unifying framework composed by reconceptualization, transdisciplinary methods and approaches, data transformation tools and platforms that support online learning and data co-curation with the goal of informing environmental policy and governance. To this end, we have identified following three areas where the issue of framing challenge is particularly relevant:

- Scale versus governance boundary in conducting research (re-conceptualization)
- Information gaps in public service delivery planning (methods and approaches)
- Co-curation of sustainability research: the role of place-based observatory and boundary organizations (data transformation tools and platforms)

Shifting focus: physical scale versus governance boundary in conducting research

Scale-oriented approach has conventionally earned a dominant place, especially in environmental research focussed on water, forestry and land and soil resources (Gregersen et al., 1989; Pahl-Wostl et al., 2021). Scale in the context of water, for example, is defined by the physical scale such as watershed, river basin and farm

[7] Biophysical analysis tends to focus on issues of resource security (see Sadoff et al., 2020).

plot. The analysis based on scale typically focuses on the biophysical aspect of the selected watershed, river basin or farm plot. But research on agricultural research for development has highlighted the role of institutions (understood here as rules) to emphasize the risks that could arise from a failure to incorporate those perspectives in environmental research. Environmental research undertaken in a disconnected and siloed manner promotes developmental interventions that are lacking in horizontal and parallel connectivity. But the Bonn Conference of 2011 sought to address the limitations of an overemphasis on a scale-oriented approach by shifting the focus of the international policy discourse from integrated water resources management to the nexus approach, thereby demanding more rigorous engagement with the governance dimension in environmental research.

Boundary-oriented approach touches upon the dynamics of governance—government institutions at different levels (central, provincial and local) to manage environmental resources. Contrary to what is believed, it is usually easier to craft operational rules (for example, to manage an irrigation system or forest) to deal with immediate and smaller-scale environmental challenges of soil erosion or water scarcity. The ability of institutions to make impact on a larger scale and in a sustained fashion is, however, dependent on how policy interventions at a lower order align with higher-order constitutional and collective choice rules. This tension between higher- and lower-order rules exemplifies how water, energy and food resources are managed by large bureaucracies.

Governments have a tendency to centralize the production and distribution of water (through dams and irrigation canals), energy (hydropower) and food (procurement for storage and distribution via warehouses) to legitimize their power. This may be evident from a study of constitutional choice rules that reflect allocation norms (shared values) that are determined by a combination of party priorities, consumer preferences and type of electoral and administrative systems (Boyne, 1996). Here, a wide range of operational norms are at play to define the scope and nature of policy choices. Allocation norms can determine how much water or energy resources can be shared between industry, agriculture and water supply (see Fig. 1.5). Similarly, coordination norms that operate at the level of collective choice rules can determine how central fiscal transfers are to be disbursed, criteria for monitoring water quality and service delivery standards that need to be met for local governments and departments to avail of taxes and transfers. Finally, equity norms that typically function at the level of operational rules can determine service delivery standards and distributional criteria for levy of tariffs on consumers in a village, ward or city.

Scale as an approach has often been the subject of focus in discussions of the WEF nexus. The focus on scale has had limited merit in distinguishing the nexus approach from IWRM besides placing water at the centre of the discussion and overlooking the critical role of environmental policy and governance. Any attempt to amplify the policy orientation of research must pivot towards laying the groundwork for a better appreciation of the role of intersectionality, trade-offs and critical mass that very often operate at the boundary of physical and institutional systems. A focus

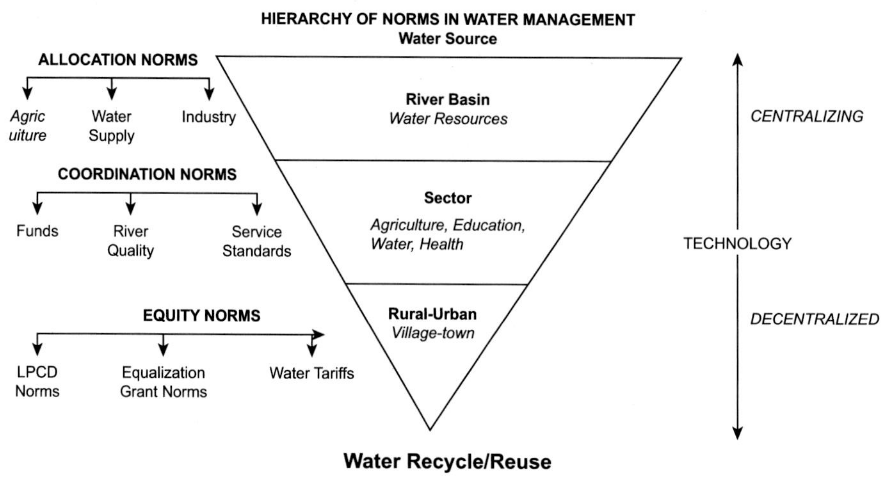

FIGURE 1.5

Hierarchy of norms in water management.

on scale as against boundary has the tendency to underplay the role of territory, authority and property in crafting rules for managing trade-offs and achieving a critical mass of networked individuals with resources and information that enable them to advance synergies in the pursuit of sustainable development. For example, if the policy goal is to achieve food security nationally, a combination of factors must come together at the city, provincial and county levels for the policy goal to be achieved. These factors could include financing, technology, trained staff and/or information about market conditions, improved seeds or fertilizer. The extent, duration and intensity with which these factors come together to support achievement of the policy goal need not be uniform, but the threshold level at which they come together to produce outcomes in support of the policy goal is referred to as the critical mass.

When policy choices are short-sighted and overlook the imperatives of duration, intensity and frequency of sustained policy support, then environmental and social outcomes that arise from policy intervention tend to lead to unintended consequences with repressive implications for environmental sustainability and management. In this connection, Elinor Ostrom in her earliest formulation of her theoretical accounts on governing the commons pointed out that some of the critical resources on which we rely upon to provide critical WEF-related services are common property resources—characterized by nonexcludability and its subtractable nature (Ostrom, 1990). The added benefit of combining the commons framework with the framework of public finance is that we begin to appreciate how sustained local level collective action is reliant upon alignment with higher-order institutional rules.

For example, when rules relating to public budgeting are unable to set aside critical resources for recurrent expenditures that are necessary for staffing and operation and maintenance of water and energy infrastructure, the planetary limits of common property resources may be stretched to their limit. In other words, forests may not be adequately safeguarded from overexploitation, rivers may be overpolluted, livestock grazing areas may become overstocked and groundwater aquifers may be pumped dry. This trend of overexploitation can transform common-property systems into open-access resources and make them susceptible to degradation with adverse consequences for sustainable development. This is a classic example of problem framing challenges that confront the scientific community in terms of designing and implementing policy-relevant research. Whereas it will be discussed more in detail in Chapter 2, in simple terms, the pyramid of governance consists of three layers: constitutional choice rules (top), collective choice rules (middle) and operational rules (bottom). While scholars in public financing tend to focus on the top two layers, analysis by common property scholars is drawn only from findings in the bottom layer of the operational rules, producing two separate world views of environmental governance that are hard to reconcile. In pursuit of more effective environmental policy, how can we then bring these scientific approaches together as complementary rather than confrontational?

How to handle missing information: Environmental planning, policy and delivery of public services

In Fig. 1.4, we indicated that a key focus of central authorities responsible for monitoring the impact of central transfers is information. Information is usually not disaggregated, frequent and reliable. To take the example, of SDGs 2, 6 and 13 dealing with targets of poverty reduction, water and sanitation, and climate change, information that is collected tends to be disproportionately skewed towards data about rainfall, water quality, precipitation or temperature with little effort devoted to understanding the institutional dimensions of trade-offs associated with infrastructure costs and pricing and consumer feedback about the quality, coverage, and reliability of public services such as water supply, irrigation or wastewater treatment. This disproportionate focus on the biophysical domain can lead us to overemphasize environmental risk and become overly optimistic about the role of technology and financing in advancing sustainability (Kurian et al., 2019). This is because models that rely on indicators that are weighted heavily towards biophysical factors tend to only highlight issues of efficiency in management of wastewater or irrigation systems, i.e., how to save energy or optimize water use through reuse. These types of analysis, however, tend to overlook more policy-relevant issues of equity, i.e., who can afford to pay for reuse services and how can those who are not covered be provided access through innovative financing schemes.

A big framing challenge for sustainability research in the context of food security for example is learning how to couple global models with local models of food production as it has the potential to encompass several key sustainability goals of equity, institutional resilience, and resource reuse. Downscaling, on the other hand, essentially refers to the process by which global models of rainfall or temperature, for example, are made more useful for use by policy-makers or utility managers at the local or regional level through a process of model validation and data valorization. The process of valorizing data contained in downscaled models could involve scientific inputs in the form of development of prototype composite models of food resilience, pilot-testing specific policy instruments based on typologies of trade-offs that have be developed through longitudinal case studies to amplify the effects of public policy. In this context, we have previously been encouraged by the examples of community score cards in South Africa, the Water Point Mapping project in Tanzania and use of remote sensing to identify viable groundwater management units in India (Turral and Kurian, 2010). A key feature of downscaled and coupled models of food risk would be an explicit attempt to frame the challenge in terms of trade-offs rather than in terms of finding perfect solutions to environmental resource management problems whose links and consequences for public services are only partially understood. Holistic solutions would demand that we focus on governments as key actors to be able to appreciate the fact that institutional analysis has much to offer in terms of effecting change at scale, rather than be limited by the guidance offered by dispersed field trials and resource management case studies that are constrained by the specific conditions of the site and the statute of limitations imposed by the period of data collection. While we highlight the significance of engaging governments it is important to emphasize that are not against community level initiative; only that policy support is key to scaling up and scaling out the results of small scale success for which reform of institutional structures and processes involving governments in large measure is key.

Paradigm shift in scientific practice

A key contention of this volume is that sustainability risks have been conveniently posed in terms of security risks, but the specifics of boundary conditions have been overlooked (Sadoff et al., 2020). For example, discussions about the WEF nexus have been posed in terms of increased risk of a higher water footprint of different energy portfolios, a higher water footprint of agriculture, energy intensification of food transport or energy intensification arising from demand for desalination and water reuse (Scott et al., 2015). While such descriptions maybe true, an inability to engage with institutional context makes it difficult to frame global prescriptions about pathways of adapting to climate change in terms that make the search for solutions relevant to regional political economy considerations. Furthermore, the solutions to problems have been posed in the tradition of rational actors, making rational choices when information is made available to them (Perry and Easter, 2004). But when viewed from the framework of trade-offs operating in coupled human–environment systems that are characterized by unpredictable behaviour of agencies and agents, it should be apparent that political economy considerations may play an important role in environmental decision-making in addition to considerations of externality and transaction costs of actors who are presumably acting rationally.

The July 26 issue of The New York times reported on attempts to undertake climate modelling to predict global migration hot spots. The article referred to a study by Kreuger and Oppenheimer published in the 2010 proceedings of the National Academy of Sciences that concluded that Mexican migration to the United States spiked during periods of drought. Several studies that have followed in the wake of the Kreuger and Oppenheimer application of econometric modelling to the climate migration challenge have concluded that while climate is rarely a main cause of migration, it is almost always an exacerbating one. The studies have also been able to map out from where migrants are likely to move in the future as a result of changes in the climate—sub-Saharan Africa, South Asia, and South America were key migration hot spots driven by droughts while the Mekong delta was a key hot spot where floods are a key driver. However, such studies have been unable to address the key triggers influencing people to migrate—for example, declining groundwater.

To overcome this shortcoming the same article reported that the World Bank has begun addressing the specifics of climate-induced migration by gathering data sets of political stability, agricultural productivity, food stress, water availability, social networks and weather all focused on predicting human decision-making. Then with the objective of identifying the specific thresholds of climate-induced migration, the research went beyond merely guessing what an individual will do and multiplying that decision by the number of people in similar circumstances. Instead, the modelling effort looked across entire populations, averaging out trends in community decision-making based on established patterns and then seeing how those trends play out in different scenarios. This attempt to engage with real-world conditions resulted in 10 million data points which only a supercomputer housed at the National Centre for Atmospheric Research in Wyoming could process efficiently.

The aforementioned example of modeling climate-induced migration illustrates two reasons why there is the need for a reform of scientific practices. First, it is related to the mismatch between mechanism of research and architecture of environmental planning. Second, decision support tools are primarily statistical modelling analysis based on insufficient and/or narrowly defined data, which typically bear scoping challenges as we described earlier. In this connection, hyperlocalized case studies are often adopted to compensate for the numeric generalization. While case studies can provide rich anecdotal evidence, they fall short in terms of generating generalizable principles tested against local specificities that can guide sound decision-making in environmental policy and planning.

In sum, the scale, frequency and intensity of climate-induced environmental and social change necessitates a paradigm shift in the way in which we engage with sustainability research so as to comprehend the wider and more complex landscapes of political economy to develop a nuanced understanding of the world around us. We will argue in this volume that greater attention to the framing problem in research and complimentary measures—creation of linked databases, online learning, data valorization and data visualization can go a long way in enhancing the policy orientation of research (Pearl and Mackanzie, 2018).

Place-based observatories—a tool for enhancing the policy orientation of sustainability research

As can be ascertained from the above discussion sustainability research must be based on multi-dimensional indicators that need to be monitored over time. While a "synthetic index" that is composed of socio-economic, institutional and bio-physical indicators may be useful there are several innovations that need to be undertaken. First, it is important to distinguish the various deciles and percentiles of relevant indicators. Second, the synthetic index needs to be connected to policy via institutional instruments- guidelines, directives, notifications, standards and circulars. Third, data that is needed to test the impact of institutional instruments on public policy outcomes need to be ascertained via data that is co-curated. For these three reasons place-based observatories are critical to facilitating a paradigm shift in sustainability research. We have previously identified attempts by UNHABITAT to set up the Global Urban Observatory and by WHO to establish Global Observatory on Health Research and Development that indicate encouraging signs of change (Kurian et al., 2016). The idea of gravitating towards relying upon place-based observatories is founded on a few core principles (Hall and Tiropanis, 2012):

- Access to distributed repositories of data, online social network data and web archive
- Harmonized access to distributed (linked) repositories of visual/analytical databases and tools to support a variety of quantitative and qualitative analysis
- Shared methodologies that facilitate harvesting of additional data sources and development of novel analytical methods and visualization tools
- A forum for discussion and online learning about an ethics framework on archiving and processing of web data and related policies
- A data licensing framework for archived data and results of processing those data

In this connection, we have previously reported on our success with employing place-based observatories[8] to guide our work on developing and pilot-testing a monitoring methodology for SDG 6.3 (Kurian, 2020). We find place-based observatories enable us to test critical hypotheses for their effects on environmental and institutional outcomes. For example, we can start by examining whether political decentralization of power can be a useful strategy in mitigating the effects of droughts and floods by bringing decision-makers closer to people affected by such hazards. Second, we can also examine whether community participation in the design, implementation and evaluation of development programs can improve the resilience of food systems through grant of greater autonomy to local authorities. This is a highly policy-relevant concern as improved resilience could result in enhanced agricultural yields given the proximity of consumers to decision-making

[8] The United Nations University in Germany launched the Nexus Observatory online platform in collaboration with GIZ, Bonn, at the Dresden Nexus Conference in May 2017. For more information on the Nexus Observatory initiative, please visit https://nexusobservatory.flores.unu.edu/.

processes relating to the choice of seeds, fertilizers and information about prices of crops and inputs. Concerns related to technical improvements in agricultural production would lead us to our next logical inquiry—the state of intra/interhousehold welfare, which is the third point. Does decentralization also improve nutritional security by addressing critical issues of power and hierarchy that operate along axis of gender, class and ethnicity within and across households? (Kurian et al., 2016).

5. What are boundary organizations?- Their architecture and scientific roles

Boundary organizations typically refer to think tanks that undertake public goods research that look beyond the short-term policy horizon of sitting governments, and towards a medium-term time frame within which their analysis and ideas would have a better chance of being considered by those in office.

> *US think tanks have focused on reforming procedural dimensions of policy making in order to influence change. A 1974 study by the Brookings Institution for example, proposed the creation of the Congressional Budget Office while another study by the Centre for Strategic and International Studies (CSIS) led to the Goldwater-Nichols Act reforming the command of the US military. Recently, universities have also moved aggressively into the think tank space for reasons ranging from a desire to improve the quality of their resident expertise, the opportunity for faculty to gain additional recognition and visibility from contributing to policy projects; and the potential for this recognition to help attract more students and more public and private funding for research with potential to have real-life impacts*
>
> **Niblett, R. (2018). Rediscovering a sense of purpose: the challenge of western think tanks. Int. Affairs 94: 1409–1430.**

In a similar line, with specific reference to discourse surrounding the SDGs and the WEF several think tanks including the Consultative Group on International Agriculture Research (CGIAR), United Nations University, Stockholm Environment Institute and Future Earth have made important strides (Cvitanovic et al., 2018). Against this background based on the success of the Baltic Eye project (Cvitanovic et al., 2018), we were able to highlight four features of boundary organizations that enhance impact on policy and practice:

- the inclusion of policy analysis within diverse teams,
- the establishment of clear goals and metrics for measuring impact (publications or others such as training and design of policy instruments),
- the presence of effective leaders and
- secured funding.

Effective leaders essentially possessed skills to undertake boundary-spanning roles as follows:

- policy-scanning to identify policy windows and the emerging science needs of decision makers—for example, what lessons can be drawn from the Brooking Institutions 1974 study?
- identifying the most appropriate means to influence policy or practice—would case studies that rely on primary data always suffice?
- establishing networks among team members and external stakeholders—such as those drawn from groups of consumers, private sector utility providers and decision makers.

These are factors that need serious consideration given the fact that the core purpose of think tanks was to infuse public debate with analysis based on facts and expertise, not on opinion or bias is more important today, given the rise of big data and social media platforms. (Picketty, (2020)) rightly point out "those who believe that we will one day be able to rely on a mathematical formula, algorithm or econometric model to determine what is socially feasible are destined to be disappointed. Only open, democratic deliberation, conducted in plain natural language can promise the level of nuance and subtlety necessary to make choices of such magnitude".

We contend that the ensuing discussion will clarify that boundary organizations are best positioned to take on the framing challenge by addressing boundary conditions, public service delivery considerations and co-curation of sustainability research. To this end, specifically, with reference to environmental research, boundary organizations would need to broaden the scope of their research methodologies so as to incorporate the ways in which individuals acting in concert have the potential to be more powerful than institutions in responding to complex policy challenges, such as climate change, resource over consumption and rising costs of monitoring impacts on environmental and public health. This would necessitate investing in new research tools, from online surveys to the use of big data to underpin research conclusions. Conventional case studies or meta analysis of common property resource management will not suffice. Boundary organizations will have to additionally consider the role of agent-based modelling and nudge studies to improve the efficacy of randomized control trials.

Engaging with boundary conditions in sustainability research would entail hiring professionals with boundary-spanning skills that would enable boundary organizations to offer governments with the evidence required to prepare, predict and recover from extreme trade-offs in coupled human−environment systems. The framing of modelling exercises would need to be informed by the need for information and monitoring of policy impact for which data valorization is key. The development of typologies of trade-offs, pilot-testing of prototype models of environmental decision-making and establishment for platforms for online learning and co-curation and co-design of linked databases and data valorization would enable continuous engagement with policy questions and challenges of capacity development. In short what is needed is not more models—but a model of models—one that improves our capacity for social learning and data interpretation about the institutional trajectories of developmental interventions and how some have succeeded in improving the human condition while others failed despite recourse to technology and financing.

6. Organization of the book

The overarching aim of this book is to advance a theory of change on public action on environmental governance by examining two key developmental concerns. The first relates to the role of case studies: the project will explore the groundwork needed for integrating remote sensing and online learning tools to amplify the results of RCTs in policy research. The second concern relates to modelling of the SDGs: the project will explore the use of a Nexus framework to identify complimentary measures that can connect scientific modelling with the policy process.

This book we hope will outline the broad contours of what we call *boundary science*—an emergent field (encompassing theory, method, models and data transformation techniques) with the potential to offer improved explanation and bridge the gaps between science and policy in the context of international development. We define boundary science as theorizing that aims to elaborate upon the role of boundary organizations that address the conditions of territory, authority, and property by paying attention to the state of knowledge creation and translation that can potentially be advanced by an expansion of boundary-spanning roles in sustainability research projects. Chapter 2 of the volume will outline a theoretical framework for boundary science by elaborating upon the concept of blind spots in environmental governance. In this context, discussions evolve around key theoretical developments related to three pillars of environmental governance—norms, institutions and organizations while highlighting shifting changes in mainstream environmental discourses. Chapter 3 will unfold the methodological constructs of boundary science by engaging with issues involved in undertaking analysis of institutional trajectories in the context of environment and development nexus.

Chapter 4 will discuss some of the applications of boundary science research for capacity building that aids co-curation of data and models and co-design of research inquires, which promotes a more holistic understanding of environmental policy and governance. In this connection, the applications of open-access platforms that are supported by information and communication technology (ICT), virtual reality (VR), artificial intelligence (AI), remote sensing tools and big data analytics will be examined. This chapter concludes by investigating the versatility of boundary science in connection to environmental back-casting to explore opportunities and challenges for future research and policy intervention.

Useful links

Belmont Forum
Brundtland Report
Club of Rome Report
Hyogo Protocols
Sendai Framework of Action
Paris Climate Accord
Global City Indicators Facility
International Forestry Resources and Institutions Research Program
Consultative Group on International Agriculture Research
Future Earth
United Nations University
United Nations Sustainable Development Goals

References

Batley, R., 2004. The politics of service delivery reform. Dev. Change 35, 31–56.

Belcher, B., Claus, R., Davel, R., Ramirez, L., 2019. Linking transdisciplinary research characteristics and quality to effectiveness: a comparative analysis of five research for development projects. Environ. Sci. Pol. 101, 192–203.

Bjorkman, J., 2010. Budget support to local government: theory and practice. In: Kurian, McCarney (Eds.), Peri-urban Water and Sanitation Services: Policy, Planning and Method. Springer, Dordrecht, pp. 171–192.

Boyne, G., 1996. Competition and local government: a public choice perspective. Urban Stud. 33 (4-5), 703–721. Nos.

Cvitanovic, C., Lof, M., Norstrom, A., Reed, M., 2018. Building University-based boundary organizations that facilitate impacts on environmental policy and practice. PloS One 13 (9), e0203752.

Dietz, T., 2010. Climate based risks in cities. In: Kurian, M., McCarney, P. (Eds.), Peri-urban Water and Sanitation Services: Policy, Planning and Method. Springer, Dordrecht.

Giz, &ICLEI., 2014. Operationalizing the Urban Nexus – towards Resource-Efficient and Integrated Cities and Metropolitan Regions. BMZ, Bonn.

Gregersen, H., Draper, S., Dieter, E., 1989. People and Forests: EDI Seminar Series. The World Bank, Washington DC.

Gregory, R., 1997. Political rationality or incrementalism? In: Hill, M. (Ed.), The Policy Process a Reader. Prentice Hill, Essex, pp. 175–191.

Hall, Tiropanis, 2012. Web evolution and web science. Comput. Network. 56, 3859–3865.

Hill, M., 2006. Development as empowerment. Capabilities, Freedom and Equality: Amartya Sen's work from a Gender Perspective, vol. 1. Oxford University Press, New Delhi, pp. 132–152.

Kabeer, N., 1994. The emergence of women as a constituency in development. Reversed Realities: Gender Hierachies in Development Thought, 1st. Verso, London, pp. 1–10.

Kurian, M., 2010. Institutions and economic development- a framework for water services. In: Kurian, M., McCarney, P. (Eds.), Peri-urban water and Sanitation services- policy, planning and methods. Dordrecht, Springer.

Kurian, M., Ardakanian, R., 2015. In: Governing the Nexus- Water, Soil and Waste Resources Considering Global Change. Springer- UNU, Switzerland.

Kurian, M., Reddy, V.R., Dietz, T., Brdjanovic, D., 2013. Wastewater reuse for peri-urban agriculture- A viable option for adaptive water management? Sustain. Sci. 8 (1), 47–59.

Kurian, M., Veiga, L., Ardakanian, R., Meyer, K., 2016. Resources, Services and Risks- How Can Data Observatories Bridge the Science- Policy Divide in Environmental Governance? Springer Briefs, Switzerland.

Kurian, M., Portney, K., Rappold, G., Hannibal, B., Gebrechorkos, S., 2018. Governance of the water-energy-food nexus: a social network analysis to understanding agency behaviour. In: Huelsmann, S., Ardakanian, R. (Eds.), Managing Water, Soil and Waste Resources to Achieve Sustainable Development Goals. Springer, Cham, pp. 125–147.

Kurian, M., Reddy, V., Scott, C., Alabaster, G., Nardocci, A., Portney, K., Boer, R., Hannibal, B., 2019. One swallow does not make a summer- siloes, trade-offs and synergies in the water-energy-food nexus. Front. Environ. Sci. 7 (32), 1–17. Special Issue on "Achieving Water-Energy-Food Nexus Sustainability- a Science and Data Need or a Need for Integrated Public Policy?" (Editors: Rabi Mohtar, Jillian Cox and Richard Lawford).

Kurian, M., 2020. Monitoring versus modelling of water-energy-food interactions: how place-based observatories can inform research for sustainable development. Curr. Opin. Environ. Sustain. 44, 35–41. Special Issue on Resilience and Complexity: Frameworks and Models to Capture Socio-Ecological Interactions (Editors: Christopher Scott and Francois-Michel Le Tourneau), Elsevier.

Mannschatz, T., Buchroithner, M., Hulsmann, S., 2015. Visualization of water services in Africa: data applications for water governance. In: Kurian, M., Ardakanian, R. (Eds.), Governing the Nexus: Water, Soil and Waste Resources Considering Global Change. Springer, Dordrecht.

Meyer, K., Kurian, M., 2017. The role of international cooperation in operationalizing the nexus: emerging lessons of the nexus observatory. In: Abdul Salam, P., Shreshta, S., Pandey, V., Kumar, A. (Eds.), Water-Energy-Food Nexus: Principles and Practices, Geophysical Monograph 229, first ed. American Geophysical Union, John Wiley and Sons.

Niblett, R., 2018. Rediscovering a sense of purpose: the challenge of western think tanks. Int. Affair. 94, 1409–1430.

Ostrom, E., 1990. Governing the Commons: The Evolution of Institutions for Collective Action. University Press Princeton, Cambridge.

Pahl-Wostl, C., Gorris, P., Jager, N., Koch, L., Lebel, L., Stein, C., Venghaus, S., Witchanachchi, S., 2021. Scale-related governance challenges in the WEF Nexus: towards a diagnostic approach. Sustain. Sci. 16, 615–629.

Pearl, J., Mackanzie, D., 2018. The Book of Why: The New Science of Cause and Effect. Allen Lane, London.

Perry, J., Easter, W., 2004. Resolving the scale incompatibility dilemma in river basin management. Water Resour. Res. 40, WO8S06.

Picketty, T., 2020. Introduction. Capital and Ideology, 1st. Belknap, Harvard University, pp. 1–47.

Pollitt, C., Bouckeart, G., 2000. Public Sector Management: A Comparative Analysis. University Press Oxford, Oxford.

Reddy, R., Kurian, M., 2015. Life-cycle cost analysis of infrastructure projects. In: Kuran, M., Ardakanian, R. (Eds.), Governing the Nexus- Water, Soil and Waste Resources Considering Global Change. Springer, Dordrecht.
Renkow, M., 2018. "A Reflection on Impact and Influence of CGIAR PolicyOriented Research," in Standing Panel on Impact Assessment (SPIA). CGIAR Independent Science and Partnership Council (ISPC), Rome), p. 34.
Sadoff, C., Grey, D., Borgomeo, E., 2020. Water Security, Oxford Research Encylopedia of Environmental Science. Oxford University Press, USA.
Scott, C., Kurian, M., Westcoat, J., 2015. The water- energy- food nexus- enhancing adaptive capacity to complex global challenges. In: Kurian, Ardakanian (Eds.), Governing the Nexus- Water, Soil and Waste Resources Considering Global Change. Springer- UNU, Dordrecht, pp. 15–38.
Turral, H., Kurian, M., 2010. Information's role in adaptive groundwater management. In: Kurian, McCarney (Eds.), Peri-urban Water and Sanitation Services: Policy, Planning and Method. Springer, Dordrecht, pp. 171–192.
Twisa, S., Kazumba, S., Kurian, M., Buchroithner, M., 2020. Evaluating and predicting the effects of land use change on hydrology in wami river basin, Tanzania. Hydrology 7, 17.
Veiga, L., Kurian, M., Ardakanian, R. (Eds.), 2015. In: Inter-governmental Fiscal Relations- Questions of Accountability and Autonomy. Springer Briefs- UNU, Dordrecht.
WSP, 2009. Urban Sanitation in Indonesia: Planning for Progress. Water and Sanitation Program, Jakarta.

CHAPTER

Blind spots in environmental governance

2

1. Introduction – Why randomized control trials are popular yet will fall short in terms of improving our understanding of environmental decision-making?

At present, natural resource management (NRM) research is dominated by biophysical perspectives of environmental change. We concur with Albrecht et al. (2018) who have argued previously that the absence of integrative analysis incorporating perspectives on constraints and opportunities from the institutional domain leaves us with an incomplete understanding of prospects for environmental management. This incomplete view can lead us to overemphasize environmental risk and be overly optimistic about the role of technology and financing in advancing sustainability. The implication of this perspective is that more information would be needed about the processes and needs of decision-making at all levels, before monitoring and evaluation can be better targeted to better support them (Waylen et al., 2019: 381).

Our analysis leads us to believe in the need for a renewed theory of change on public action in environmental governance by adapting hypothesis and explanation to insights gleaned from data and without being ambitious about fitting data to dominant models of environmental change (see also Pearl and Mackanzie, 2018). Such a renewal in scientific approach has implications broadly for how we structure learning and capacity development to inform feedback into governance structures and processes. One of the specific ways in which feedback into governance processes can be beneficial is to improve design of randomized control trials (RCTs). While there have been many RCTs looking at the performance of these technologies where the unit of randomization is the plot, there is a serious dearth of RCTs looking at randomization at the village/municipal ward or individual level – the only research designs capable of rigorously uncovering the exact causal pathways between adoption of technical options and impact on water, energy and food security (Banerjee and Duflo, 2011).

It is against this background that this chapter will provide a normative framework for integrating scientific inputs into environmental policy and planning by undertaking a theoretical exploration of three key concepts: norms, institutions and organizations which bear mutually reinforcing effects. In principle, norms are shared values, while institutions are the set of rules reflected in legal and policy instruments such as directives, guidelines, notifications, circulars and standards. Organizations

Boundary Science. https://doi.org/10.1016/B978-0-323-88473-0.00002-6

on the other hand are the public agencies that are responsible for implementation of rules. When normative change alters significantly it can have the capacity to knock institutional evolution and organizational behaviour off a previous trajectory of incremental change and make paradigm shift a distinct necessity (Kurian, 2017). We have hypothesized previously in this regard that environmental disasters or hazards such as droughts, floods or tsunamis may play a role in fostering a paradigm shift in institutional evolution (Boserup, 1990).

Furthermore, institutions have two faces: one is related to legislation and policy directives – we call this institutional environment, and the other is related to institutional arrangements. Institutional arrangements refer to aspects of government capacity that serve to effectively translate legislation and policy directives via a focus on institutional arrangements such as extension agencies who are responsible for dispersion of improved seeds, fertilizers or information of improved technology or management methods in agriculture.

For easier reference, let us further examine these concepts with an example from outside of the NRM domain. Here, we choose the issue of trafficking in women and girls, for instance. Generally speaking, many agree that trafficking in women and girls is inhumane and should not be tolerated; it should be eliminated by the government's efforts. These shared values are considered 'norms' of the population. This may be reflected in the central legislature passing a human trafficking bill which is composed of anti-prostitution law and violence against women act together with issuing a directive on inter-agency collaboration to eliminate human trafficking (institutional environment). Furthermore, the establishment of a government-owned women's shelter and focal point within relevant ministries are some examples of 'institutional arrangements' that can promote an institutional environment that discourages human trafficking. This set of rules are 'institutions' and should not be mistaken for the brick and mortar government offices. Finally, law enforcement authorities and government-mandated NGOs are the ones who are actually involved in implementing the government policy directives and legislation. They are referred to as 'organizations'.

In sum, we hope to demonstrate in this chapter how the theoretical evolution of these concepts of norms, institutions and organizations interacts with mainstream paradigms and underpins the development of the institutional trajectories assessment framework. The methodological insights that emerge from this exercise highlight the role of boundary science in effectively understanding the impetus or lack of thereof public policy responses to emerging pressures and trade-offs in environmental governance.

2. Blind spots in environmental governance: Institutional fragmentation and the poverty-environment Nexus

Environmental science literature often highlights 'hot spots' or 'bright spots' to signify the heightened biophysical risk and/or opportunities, respectively, of optimizing resource use efficiency (Daher et al., 2018). Social scientists engaged in environmental studies on the other hand have begun to voice their concerns about how little attention has been devoted to understanding the role of 'blind spots' especially given their potential to emphasize accountability and synergies in environmental governance. At a glance, these blind spots inform us that there are several guiding assumptions that shape the construction of models, the design of experimental research and the monitoring of extreme environmental trade-offs. If you examine them closely, you will also realize that certain norms, property rights and incentives for organizational feedback, learning and reform play a vital role in shaping these guiding assumptions (see Table 2.1). Table 2.1 maps out the internal logic of one of the popular unidimentional models in this regard.

Through the experience of establishing a research program at the United Nations University (UNU) that informed the pilot-testing of a methodology for monitoring the Sustainable Development Goals (SDGs), we realized the significance of applying the nexus framework to support policy-relevant research on water—energy—food (WEF) interactions. To arrive at these convictions, we carefully reflected upon our previous research that applied the conventional approach — field-based case studies, data and models to examine how 'policy blind spots' can be overcome in environmental governance in the contexts of economic development in Asia, Middle East and African regions and the design of recent innovation contests (such as the Belmont Forum/National Science Foundation, United States grant proposal).

Table 2.1 Guiding assumptions on blind spots - example of unidimensional modelling.

Issues	Models	Design	Management Information System (MIS)
Guiding assumptions	Systems model	- Data on WEF interactions heightens chances of better decision making through information sharing and coordination.	Data collected from stakeholder consultations will lead to coherent decisions
Norms-incentives-property rights	None addressed	- links to discussion of technical and management models missing	Acknowledgement of organizational mapping of feedback loops/ administration rules within decision making structures/ processes missing
Examples	WEF Nexus Tool 2.0	- Characterization of environmental challenge missing - links to regional and global environmental/ decision making models missing	Mapping of resource use hotspots

Source: (Daher et al., 2018).

We have identified following five blind spots where the benefits of applying boundary science are foremost in relation to environmental governance analysis with an explicit focus on WEF interactions:

- The divergence in decisions based on guidance offered by global versus regional environmental models: how do differences in norms at the level of international organizations, states and regional governments affect the extent of this divergence?
- The divergence in decisions based on guidance offered by unidimensional versus multidimensional models: how do differences in scientific norms of research and policy think tanks influence the extent of this divergence?
- The scale and impact of decisions based on guidance offered by RCTs of technical options and management models: how do differences in institutional environment affect the uptake of results of RCTs in decision-making?
- Assumptions about data availability that guide the design of Management Information Systems (MIS) of WEF interventions: how do sectoral and siloed thinking that shape institutional design affect the design of MIS?
- Assumptions that guide the nature of feedback loops between MIS and decision making that supports the prediction, response and recovery from extreme environmental trade-offs: how do differences in organizational staffing, skills and training affect the robustness of feedback loops between scientific assessments and environmental decision-making?

In sum, to engage in sound analysis of environmental governance, it is critical to understand the mechanism of blind spots – how they emerge, under what conditions and contexts they are sustained and why they take time to overcome – from the perspective of norms, institutions and organizations, which will be the focus of this chapter. To this end, this chapter has three objectives: (1) to outline a theoretical framework to understand how norms in environmental governance are shaped by the evolution of state systems and scientific paradigms, (2) to discuss different institutional approaches for the study of territory, authority and property in environmental governance and (3) to contrast the different pathways of change in biophysical and institutional systems and outline their implications for feedback, learning and reform of boundary organizations.

These three objectives are drawn from the basic understanding that the features of environmental governance are defined by how states shape the construction of norms based on differences in; 1) state structure, 2) the degree of autonomy, 3) extent of coordination, 4) level of accountability, 5) form of fiscal mechanisms 6) nature of electoral systems and 7) consistency of party voting behaviour on a range of policy options. We refer to these factors of environmental governance as the 'seven boundary conditions'. We expect that by clarifying the issue of boundary conditions, we will be able to address the critical issue of institutional fragmentation, which poses major challenges for environmental governance. Causes of institutional fragmentation identified in the areas related to the structure, policy and thinking of public administration scholars and international development cooperation mirrors the flaws and limitations inherent in the conventional character of the state and the politics which fuels exclusiveness, hegemony and marginalization of sub-altern voices in the process of construction and transformation of the modern state system (Connell, 1990; Foucault, 1980; Kabeer, 1994; Kuhn and Wolpe, 1978; Mies, 1986). Here it is important to distinguish between two conceptual functions of politics which are defined as institutional (the nature of government institutions in relation to issues of rights, justice and responsibility) and instrumental (issues of power and policy (Squires, 1999: 8) In this connection, institutional fragmentation emerges as the result of competition among urban and rural government entities for central fiscal transfers, overlapping jurisdictional boundaries and incentives for management coordination among line departments and respective ministries. Too ofen, a poor understanding of the extent and causes of environmental trade-offs that emerge from the recursive effects of infrastructure costing and pricing rules can exacerbate institutional fragmentation. Improper selection of programs for financing by multilateral agencies, improper selection of technologies for water treatment and erroneous targeting of beneficiaries by development programs/projects are some common examples (Kurian, 2013; Walle and Gunewardena, 2001).

In other instances, weak feedback loops between legal and policy structures and processes, spatial and temporal variation in distribution of biophysical resources and differences in socio-economic and demographic characteristics may result in poor management outcomes. For instance, in the case of the Ford Foundation Joint Forest Management project, we found that an important assumption that guided program implementation was that if landowners were given access to irrigation, they would reduce their dependence on the forest for livestock fodder. But since feedback mechanisms were weak, it became difficult for the Forest Department to respond to the trend of declining interest in animal husbandry due to an expansion of low-cost groundwater technology and sub-division of agricultural land, thereby compromising the operation and maintenance of irrigation infrastructure (Kurian and Dietz, 2012). Similarly, in an acknowledgement of the need to improve feedback mechanisms the Dutch Water Boards have explored the use of performance benchmarking to link budgetary disbursements to achievement the policy objective of water quality and wastewater treatment (Blokland, 2010; see Table 2.2).

Keeping everyone happy: Challenges of distribution of interests and resources in the poverty-environment Nexus

One of the major drawbacks of institutional fragmentation we contend is that it drives the emergence and sustenance of blind spots in environmental governance, severely curtailing the ability of international, regional and local authorities to align their actions that would enable them to prepare, respond and recover from extreme environmental trade-offs (Kurian and Ardakanian, 2015). Furthermore, institutional fragmentation lays bare some key divides in environmental governance, reminding us that the dynamics behind public policy choices relating to infrastructure versus services, centralized versus decentralized governance structures, public versus

Table 2.2 Ranking of dutch water board.

Water Board	Treatment	Finance	Environment	Innovation	Stakeholders
Hoogheemraadschap Amstel, Gooi en Vecht		11	23	4	17
Zuiveringschap Limburg	15	3	20	15	12
Hoogheemraadschap van Delfland	24	15	27	17	
Zuiveringsschap Hollandse Eilanden en Waarden	14	5	18	6	12
Hoogheemraadschap van Rijnland	8	4	5	12	
Hoogheemraadschap Hollandse Noorderkwartier	13	20	2	1	22
Waterschap de Dommel	22	9	19	24	17
Hoogheemraadschap van West-Brabant	16	5	11	17	12
Wetterskip Fryslan	12	13	2	9	17
Hoogheemraadschap de Stichtse Rijnlanden	3	21	9	9	12
Waterschap Rijn en IJssel	10	9	8	2	3
Waterschap Rivierenland	17	19	15	7	3
Waterschap Regge en Dinkel	1	8	9	8	3
Waterschap Vallei & Eem	5	17	12	19	3
Waterschap Veluwe	23	2	23	14	3
Hoogheemraadschap van Schieland	7	18	1	27	1
Waterschap de Maaskant	2	12	7	22	1
Waterschap Hunze en Aa's	20	24	21	16	3
Waterschap De Aa	6	1	15	24	3
Waterschap Groot Salland	21	5	13	3	3
Waterschap Zuiderzeeland		24	4	23	22
Waterschap Noorderzijlvest		21	25	19	3
Waterschap Zeeuwse Eilanden	19	15	17	11	17
Waterschap Reest en Wieden	9	24	5	13	12
Waterschap Velt en Vecht	18	23	26	4	17
Waterschap Zeeuws Vlaanderen	11	24	21	19	22
Hoogheemraadschap Alm en Biesbosch	4	14	13	26	

Rank 1–9 Rank 19–27
Rank 10–18 Nodata
UvW (2003)

Source: UvW (2003) cited in Blokland 2010: 255.

private management models, short-term versus long-term planning horizons and promotion of efficiency versus equity goals are much more complex and thus need to be examined beyond recourse to simplistic binary models. The divides in environmental governance in turn highlight a key challenge confronting public policy: why do we pursue environmental management? One challenge is obvious: how can we protect against planetary boundaries being breached due to anthropogenic activity. The second is how to distribute the benefits of enhanced management among human populations in the form of ecosystem services: safe water, aquifer recharge and silt transfer.

Very often decision-makers are confronted with the challenge of making decisions that have implications for both of the above mentioned priorities at the same time: safeguarding planetary boundaries while ensuring a balance in how benefits of environmental management are distributed. This is why it is important as scientists and modellers to approach the study of public policy challenges in terms of trade-offs and to focus on identifying pathways and mechanisms with the potential to influence normative change. To focus exclusively on identifying technical options or strategies for improved management of environmental resources would limit our ability to scale out and scale up the results of global public goods research. It is in this context that it is important to appreciate the role of public investments in infrastructure in transforming ecosystem services into benefits for consumers. Let us explain this point in a more concrete manner by using water related infrastructure as an example, Irrigation systems transport water to individual farms while water treatment plants ensure that safe water is made available in the form of water supply and canal systems can transport silt to users located downstream of a river system. It is important to appreciate, however, that infrastructure such as water treatment plants costs money to establish and they can be financed from a variety of sources, both public and private. In most cases, governments are important sources of financing for delivery of public services for which recourse to taxes and tariffs is crucial. There have been initiatives that have sought to engage the private sector for which a regulatory framework is key (see Box 2.1 – Malaysia). Multilateral banks have in recent years experimented with a range of public–private financing regimes/contract types for critical water and wastewater infrastructure such as build operate and transfer (BOT) (see Box 2.2.)

What the aforementioned discussion highlights is the tension between how the benefits of infrastructure projects are distributed among different groups in a population. The benefits of infrastructure can prioritize different sections of a population based on a range of metrics: income, wealth or ethnicity. Norms in turn can specify exactly how benefits are distributed by clarifying minimum–maximum thresholds for each of the aforementioned metrics. For example, households with an income of less than 40,000 US dollars annually may pay a tariff of between 5 and 10 US dollars per cubic meter of water supply. Some scholars with a particular interest in developing countries have pointed out the dynamics of this political tension between infrastructure benefits and their distribution to demonstrate striking challenges confronting envoironmental governance in the context of the poverty–environment nexus.

Box 2.1 Private partnership in sewerage financing, Malaysia

The Malaysian programme for private participation in sewerage illustrates some of the disappointments that can occur when an aggressive private participation plan is put in place to mobilize finance and accelerate investments. After a few successful water and sanitation Build, Operate and Transfer (BOT) projects in Malaysia, the government chose to support a national sewerage project, the Indah Water Konsortium (IWK). This project arose from concerns over local government's weak technical and financial capability in the face of poorly maintained facilities and rising demand for better sewerage services. An unsolicited proposal was brought to the government and approved rapidly in 1994. Investments and the level of service improved dramatically in the immediate term. However, even before the economic crisis of 1997, consumers objected to the tariffs imposed. The tariff structure originally stipulated in the agreements was suspended without compensation for private contractors, and a new tariff structure was only established in 1997. The economic crisis then prompted further reductions, while the IWK discovered that the rehabilitation needed was more than anticipated. As a result, the government felt obliged to provide financial support to IWK, including long-term soft loans amounting MYR 450 million. This transaction could have been designed better: the economic crisis multiplied the difficulties tremendously, but the problem had already emerged before devaluation aggravated them. While private participation doubtless brought considerable implementation capacity to the task, they did not resolve the fundamental impediments to making provision of sanitation services a viable financing proposition. The government succeeded in attracting private investment, but the structure of guarantee provided and the nature of risks involved in the project were such that both the capital mobilized and the physical achievements of the projects were much less than originally expected.

Source: *Urban Infrastructure Finance from Private Operators: What Have We Learned from Recent Experience? World Bank Policy Research Working Paper 4045, Nov.2006 cited in Kurian (2010):136–137.*

Box 2.2 Contract definitions for private public partnership

- *Service contract:* Restricted provision for technical services through simple service contracts, where the public authority retains overall responsibility for operation and maintenance of the system, except for the specific, limited scope services that are contracted out
- *Management Contract:* contracted company is transferred responsibility for entire operation and maintenance of a system and is paid a fee to operate water supply and sanitation services
- *Lease Contract:* private company leases the water supply and sanitation assets and maintains and operates it, in return the right to water and sanitation revenues, while the public authority remains the sole owner of the assets
- *Concession:* Private contractor has overall responsibility for services including operation, maintenance and management as well as capital investments for expansion of services, while the public authority retains overall responsibility to regulate operations of the private operator and
- *Divesture:* Private operator is given full responsibility for operations, maintenance and investment and legal ownership of assets are transferred to the private company

Source: *Information drawn from World Bank (2009).*

Here, a brief explanation of background of poverty-environment nexus discussion may be useful. Most environmental problems, especially those relating to common property resources (water pollution, land degradation, deforestation and air pollution), are shaped by geography. Geography highlights the role of boundary conditions at the intersection of biophysical and administrative scales. Different geographical scales of analysis can highlight different typologies of trade-offs. 'In the case of air pollution for example, the theoretically appropriate scale is influenced by the dispersal characteristics of the pollutant and medium: particulate pollution from cement mills may only be dangerous in one urban region; acid rain from sulfur emissions may damage forests hundreds of miles from the source; and eutrophication from fertilizer runoff may affect fisheries a thousand miles downstream from the farms that are the source of the problem' (Dasgupta et al., 2005:618). Furthermore, if much of pollution's impact is felt far downwind or downstream, failure to find a nexus in the local data may simply mask transboundary effects. Most particulate air pollution settles locally; locally generated plumes of fecal coliform pollution generally persist for several miles downstream but would present transboundary problems only in border areas. Therefore, from the point of view of impact assessments, a criterion that highlights the significance of the nexus would be the disproportionate environmental damage in areas with high concentrations of poverty.

Against this background, the environmental degradation issues related to waste water treatment and air pollution have been recognized as a pressing policy issue in densely populated refugee camps in developing countries (Mukul, 2019). The International Organization for Migration estimates the number of environmental refugees would have swollen to 200 million by 2050 (IOM, 2015). In this respect, policy recognition of ascertaining the strength of the poverty—environment nexus perspective is crucial to guiding more effective public policy on environmental challenges but with caution. Some argue, for example that if levels of poverty and inequality among given populations are not mitigated[1] due to improved forest or water management, then forest and water management outcomes should be differentiated from the objectives of poverty reduction. While this will be the case in remote locations with small populations or large population centres with little reliance on environmental resources in close proximity for their livelihoods, in many cases the relationship between poverty and the environment can be observed in the form of more complex interconnections and interdependencies. From the point of view of policy intervention, it is important to understand that co-relation between

[1] Development policy interventions on women through efficiency and anti-poverty rhetorics have been criticized for their highly repressive and contradictory implications. See Jackson (1996) and Sen (1994) for similar accounts.

environmental damage and high poverty concentrations does not necessarily mean causality. Similar to continuing debates on whether to regard poverty as the sole causality of migration, the existing literature suggest that while poverty is identified as a significant cause of environmental degradation, it is not the sole driver of the trend. In fact, two contrasting schools on the poverty-environment degradation nexus both acknowledge the importance of taking into consideration how institutional and market failure also contributes significantly in shaping the behaviour of the economically disadvantaged group which leads to environmentally unsustainable activities (Duraiappah, 1996). It is thus vital to stress that only when the right nuance of these interdependences are appropreately captured in typologies of tradeoffs, then can public policy benefit and mitigate extreme tradeoffs by enhancing scope for synergies across different sectors involving forest or water departments and ministries.

In this connection, the literature on collective action has emphasized the role of distribution of interests and resources in determining the success of synergies for management of common property resources. Collective action literature with a particular focus on management of common property resources has for a long time been convinced that group homogeneity is good for promoting synergies. But recent analyses show that in the start-up phase of collective action, a certain degree of heterogeneity may promote synergies by making available resources in the form of time, money and leadership qualities to ensure that public infrastructure is properly operated and maintained (Jones, 2004; Kurian and Dietz, 2012). In this respect, drawing upon analysis guided by intersectionality approach (Crenshaw, 2017), collective action is considered as the outcome of the on-going process of group identity dynamics which is composed of intersection of different traits and social locations of its members while these traits shape their priviledges and vulnerabilities within the group, community and society in society[2]. The discussion also highlights the importance of continuously aligning microlevel operational rules such as irrigation tarriffs with management strategies and policy frameworks that are reflected in higher-order institutional rules that are supportive of participatory management (for example, a community-based forestry). The strength of synergies may also be advanced through boundary organizations by imparting skills and training on ways of ascertaining the poverty—environment nexus.

3. Norms: Fiscal mechanisms, electoral systems and monitoring regimes

What are norms?

In the previous chapter, we have briefly discussed the difference between scale-oriented and boundary-oriented approaches. Along this line, we will introduce the

[2] For more details on the intersectionality approach please refer to: Crenshaw (2017).

concept of boundary science by elaborating upon the perspectives emerging from our discussion on boundary-oriented approaches. For this purpose, we will borrow from the literature covering disciplines of geography, political economy, public finance and NRM. Theoretically, boundary science draws upon two binary elements of analysis in environmental science: biophysical scale and the administrative scale. Biophysical scale refers to boundaries defined by physical attributes such as watersheds, forest catchments and underground aquifers. Administrative scale refers to the coverage area for delivery of specific public services such as water supply, wastewater treatment or irrigation. Depending on institutional context (degree of vertical or horizontal dispersion of political power which we referred to in Chapter 1), administrative scale could be determined by boundaries of village, town, municipal ward, irrigation water user group or forest protection committees.

Scholars of environmental policy and governance often hear students and practitioners referring to farmers and consumers of water supply as stakeholders and beneficiaries of development programs. While farmers and water consumers are indeed beneficiaries, governments play an important mediatory role in the design, monitoring and evaluation of developmental interventions. When development programs, especially those designed by external donors and nongovernmental organizations (NGOs) target farmers and consumers directly, they miss out on the potential to make an impact at scale (Kurian et al., 2016). Furthermore, as we pointed out in Chapter 1, when public budgets maintain several items 'off-line', accountability can be seriously undermined in the way in which taxes and tariffs are appropriated and spent to address development priorities.

As demonstrated through several initiatives by the World Bank, however, collaborating with governments may ensure better accountability while increasing the pool of potential beneficiaries of development (Gittinger, 1982). This is attributed to the role of government as the catalyst of local operatives and this will provide a rationale for building capacity of government agencies as a matter of priority. Government regulation and practice enshrines local understanding of norms and rights, while government institutions are vested the ultimate power to advocate or discourage certain norms or ideas, which are external to their value system. Therefore, to understand the important role of government, it is imperative that we understand the normative frameworks that shape the boundary conditions that mediate the impact of infrastructure construction and maintenance on environmental and policy outcomes.

To begin, norms have been defined as 'shared values' (Ostrom, 1990). States play an important role in articulation of shared values such as social equality or environmental sustainability. It is important to recognize that there are structural, cultural and functional differences in the structure of states that shape the process by which shared values are articulated. There are three key characteristics of norms we must bear in mind. First, the differences within state structure create variation in the dynamics whereby norms are processed. For example, state structure may shape the rules that establish criterion (such as income, ethnicity or tax revenue) for demarcation of wards/electoral constituencies that may ultimately determine the recipients of central fiscal transfers (Abelson, 2003; Boyne, 1996). The following is a list of some key axis along which state structures may differ (Pollitt and Bouckaert, 2000):

- The constitution – a structural feature
- The nature of executive government at central level: a mixture of structural and functional features
- The nature of differences between political executives (ministers) and top civil servants: a functional element with cultural overtones
- The dominant administrative culture: referring to expectations of staff of an organization about what is 'normal' and 'acceptable'
- Degree of diversity among main channels through which ideas come that fuel public management reform: cultural and functional features

In addition to differences in state structure, norms are shaped by seven boundary conditions, which is our second point. First, the degree of autonomy (vertical dispersion of authority) – that is to what extent the different levels of government share authority. Along this dimension, there are three broad types identified. They are: (a) unitary and centralized states, b) unitary but decentralized states and (c) federal states (see Table 2.3). The second feature concerns the degree of horizontal coordination at central government level – the extent to which central executives can organize ranging from pole of highly fragmented to highly coordinated. Third, states vary in degrees of accountability – by how officials are held responsible for using appropriated budgetary funds and levying certain rates of taxes for delivery of specific public services (Allers and Ishemoi, 2011). Fourth, it is related to electorial systems- states vary by the process that is employed to ascertain consumer preferences – first past the post or winner take all electoral contests. Fifth, fiscal mechanisms- states vary by the process that is employed to aggregate consumer preferences – aggregation of priorities at ward, regional, city or national level. Sixth, party behaviour- depending on the level at which priorities of citizens get aggregated one will also have to contend with institutional practices that are particular to multi-party states. For example, party whip may enforce decorum to ensure a unified position on a policy issue. Seventh, differences in State structure. Very often, this discussion is confused by the simplistic argument about whether democracies are better than authoritarian regimes at addressing sustainability concerns.

Finally, policy is the normative expression of the will of government. While the boundary conditions discussed earlier act as important 'normative filters' of the will of government, they find expression through policy instruments and a monitoring and evaluation regime. The literature on adaptive management has emphasized the importance of designing and revising plans that respond to improved understanding of uncertainty in the biophysical realm (Holling, 1978). The adaptive management literature also highlights the need to further a systems approach and stakeholder engagement in NRM (Williams, 2011). While these aspects are important, most evaluations of public policy with regards to environmental management overlook the importance of instruments such as guidelines, notifications, circulars,

Table 2.3 Types of politico-administrative regimes.

Country	State structure	Executive government	Minister-civil servant relations	Administrative culture	Diversity of policy advice
Australia	Federal-coordinated	Majoritarian	Separate, mildly politicized	Public interest	Mainly civil service until 1980s
Canada	Federal	Majoritari an	Separate	Public interest	Mainly civil service
Sweden	Unitary, decentralized	Intermediate	Separate, increasingly politicized	Originally legalistic, changed to corporatist	Academics, trade unions
France	Unitary, formerly centralized, coordinated	Intermediate	Integrated, fairly politicized	Predominantly *Rechtssaat*	Mainly civil service
Germany	Federal, coordinated	Intermediate	Separate, fairly politicized	*Rechtssaat*	Mainly civil service (plus a few academics)
The Netherlands	Unitary, decentralized, fairly fragmented	Consensual	Separate, fairly politicized	Originally legalistic, pluralistic, consensual	Broad mixture (civil service, academics, other experts)
New Zealand	Unitary, centralized, mildly fragmented	Majoritarian (until 1996)	Separate, not politicized	Public interest	Mainly civil service
UK	Unitary, centralized coordinated	Majoritarian	Separate, not politicized	Public interest	Mainly civil service, recently think tanks, consultants
USA	Federal, fragmented	Intermediate	Separate, very politicized	Public interest	Political appointees, corporations, think tanks, consultants

Pollitt, C., and Bouckeart, G. (2000). Public Sector Management: A Comparative Analysis. Oxford: University Press Oxford. cited in (Kurian and Turral, 2010):174.

standards and directives. The degree to which the notifications are codified, implemented, monitored and evaluated would give us an idea about both the norms that undergird them but also the extent to which they are effective in achieving changes in agency and individual behaviour and possibly impact in terms of altered environmental and social outcomes. For example, norms that define poverty lines or safe or unsafe levels of water pollution can reflect differences and changes in standards. Notwithstanding, changes in policy priorities of elected governments it is important to emphasize here that higher-order norms[3] are usually more difficult to revise and alter when compared with lower-level operational norms, often necessitating that consensus be built across ideological divides.

Monitoring regimes usually rely on primary data to evaluate the effectiveness of public expenditure and environmental outcomes (Waylen et al., 2019). Public administration literature also points out that public departments and agencies usually purport to use monitoring data to justify increases in public budgetary appropriations (Kurian and McCarney, 2010). Others have pointed out in the context of discussions on the environment—development nexus that sectoral increases in budget allocations especially when there is a weak correlation between environment and development outcomes can lead to rebound effects or un-intended consequences of policy action (Dasgupta et al., 2005). In other words, sectoral budgetary strategies may make things worse than was the case previously because they encourage competition between sectoral ministries and departments. Therefore, acknowledgement of the importance of the normative basis for policy action is key to advancing synergies in environmental decision-making.

There are two implications of this analysis for the design of robust monitoring regimes: (1) most monitoring programmes collect a lot of primary data on environment and socio-economic indicators but without demonstrating a link between them as an expression of policy and (2) while there may be demonstrable links between data and updated or revised management, these links cannot be discerned from publicly available information. This is primarily because of the difficulty of articulating and defending an explanation of a particular policy direction in the context of inconsistencies and contradictions that tend to creep in as the seven boundary conditions play themselves out during the lifespan of elected governments. Furthermore, it may be possible that feedback could occur to higher levels and even inform the re-design of policies and management strategies, but again this is not normally tangible to detect using publicly available information. These findings suggest that there is an important entry point for science to make a difference by improving our understanding of the opportunities to enhance the robustness of feedback loops between environmental modelling and decision making (Kurian, 2020).

---

[3] For an interesting study of higher-order norms with respect to formulas for intergovernmental grants and their impact on extent of political influence in decision-making, see Allers and Ishemoi (2011).

4. Institutions: Water—energy—food (WEF) interactions, common property resources, agent-based modelling, social networks and the nexus framework

Why institutional analysis matters?

Global challenges such as urbanization, climate and demographic change have exacerbated in recent years. In Fig. 2.1 we reflect upon the fact that several important initiatives supported by global public goods research have been shaped by mainstream development thinking and intellectual movements. For example, de-regulation emerged as a key consideration in mainstream development thinking from 1980 onwards. This trend found support in intellectual movements that sprouted theorizing on common property resources; Institutional Analysis and Development (IAD) framework, agent-based modelling and social network analysis. In the aftermath of the Bonn conference of 2011 circular economy thinking supported global public goods research and programs on Water-Energy-Food (WEF) interactions.

Against this background, it is important to state that previous discussions on integrated water resources management (IWRM) have emphasized issues of integration through an analysis of (1) intersectoral competition for surface freshwater resources, (2) integration of water management at farm, system and basin scales, (3) conjunctive use of surface and groundwater resources and (d) prioritizing water for human consumption and environmental protection (Kurian and Turral, 2010). But others have been less optimistic of IWRM, pointing out that the approach

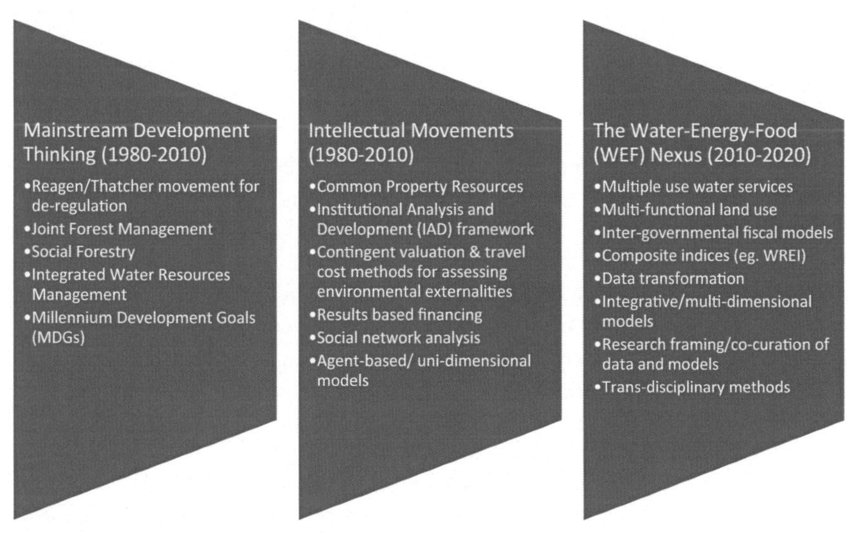

FIGURE 2.1

Trajectory of Mainstream Development Thinking and Intellectual Movements on Environmental Governance.

neglects political economy considerations through a focus on 'natural boundaries' and 'neutral planning and participation' (Wester and Warner, 2002: 65). It is precisely this critique of IWRM that the nexus approach is considered better equipped to respond to by providing an improved understanding of what measures can address trade-offs in enviromental decision making that can advance the cause of sustainable development (Kurian and Ardakanian, 2015). The critique of IWRM is well founded because it overlooks the importance of administrative boundaries. Administrative boundaries are extremely relevant from an implementation point of view since they capture the hierarchy that is implicit in authority structures that shape multilevel governance of environmental resources (Kurian, 2018). The issue of authority structures highlights the need for an institutional framework to examine sector-bound decisions characterized by poor participation mechanisms and taken with little consideration for the effects of those decisions on other sectors leading to rebound, cascading and other negative effects that are exacerbated by mutual interdependencies between resources, services and risks (Howarth and Monasterolo, 2016).

The critique of IWRM has put a spotlight on environmental policy and governance whereby the nexus approach emphasizing concepts of trade-offs and synergies has highlighted several promising lines of enquiry (Kurian, 2017) such as: (1) Why is the governance dimension important to undertake an integrative analysis of WEF challenges? (2) What does the nexus approach connote in normative and institutional terms? (3) What does implementation mean in nexus terms? (4) How can we establish if the nexus approach is an improvement over business as usual? and (5) What tools are available that would enable translation of results of scientific research to create an evidence base that would enable decision-makers to act in support of sustainable development?

What are institutions?

Contrary to the popular perception that equalizes institutions as organizations, institutions are defined as "rules" shaped by shared values and norms. As you recall, we made a distinction between constitutional, collective choice and operational rules (Ostrom, 1990) in the previous chapter. In this connection, we elaborate further on this concept of rules with regards to agricultural development by drawing upon two concepts – institutional environment (for example, constitutional rules – property rights and/or policy directives) and institutional arrangements (for example, public sector extension agencies and/or financial institutions). Some authors have pointed out that in the absence of supportive institutional environment, institutional arrangements will not flourish and vice versa.

It is important to recognize that success of institutional arrangements which will be discussed in detail later in this chapter is contingent upon the appropriate institutional environment. Dorward et al. (2005) rightly point out that 'it will not be possible to map out in advance a desirable or optimal development path for an economy, community or industry: its path will be dependent incrementally by (path dependent) political processes and trade-offs between different stakeholder groups as they act through electoral, taxation and budgetary mechanisms[4] to change or preserve the institutional environment and institutional arrangements to advance their interests'. But it is clear that the returns from investing in novel institutional arrangements in contexts where the institutional environment is poorly developed could exacerbate risks of natural shocks, price risks, economic coordination risks and risks of opportunism or rent seeking behaviour (Fan and Hazell, 2000).

There are two implications arising from this analysis. First, theoretically, it means that Neo-Classical economics in spite of its intellectual contributions to advancing our understanding of formal markets, complete information, transaction costs, externalities and rational action needs to be revisited to accommodate for the role that non-market institutions such as microfinance- and community-based NRM groups can play in substantially addressing the coordination challenge by lowering transaction costs. For example, as demonstrated in the case of adoption of community-based irrigation management, investing in innovations in institutional arrangements in dryer and poorer regions/conditions may yield greater returns than investing in changes in institutional environment, i.e., revision of property rights and land reform legislation (Fan and Hazell, 2000). Second, lessons from the pro-poor trajectory of change in institutional arrangements may subsequently inform the development of the institutional environment through proper engagement with political economy considerations of service delivery, taxes and tariffs and costs of establishing and operating infrastructure. Here the experience of fostering nested and incremental changes in land reform, local markets for seasonal finance and access to appropriate technology can provide a robust explanation of the initial success of the green revolution and how and why it began to unravel over time. The experience should also mollify our expectations that replication of success in one institutional context can be easily facilitated elsewhere via international good practice exchange. The lesson for modellers is that success is often incubated at the intersection of several supporting institutional features. Therefore, multi-dimensional models that can engage with this complexity of socio-ecological systems will be well served by typologies that review existing data, models and expert opinion to arrive at credible explanations of success and failure in environmental management.

Against this background, in this section, we will examine the theoretical trajectory of institutions in the context of agricultural development/environmental science through highlighting four main blocks: WEF interactions, common property resources (CPRs), agent-based modelling (ABM) social network analysis (SNA). Our proposed WEF nexus framework is drawn from critical analysis and synthesis of these four theoretical underpinnings which we rely upon to demonstrate institutional trajectory analysis. We intend to shed light on key shifts in theoretical discussion so as to argue that global public goods research should be focussed less on the

[4] See Boyne (1996).

conventional business of unpacking case studies but on developing typologies of trade-offs as a means of monitoring the trajectory of institutional innovation. This exercise of monitoring the trajectory of institutional innovation is likely to reveal several applications of data science that transform data to persuasively report on the effects of development interventions on vulnerable populations.

Water—energy—food interactions: Failure to capture non-monotone, non-linear and recursive effects of institutions

First there are two schools of study of WEF; WEF resource interactions and WEF Nexus. The literature on WEF resource interactions belies a focus on biophysical resources and systems analysis. WEF resources analysis has focussed on earth systems – the 'Anthropocene', an era in which human activities have significantly impacted earth system functioning (Crutzen and Stoermer, 2000). Here, Rockstrom et al. (2009) have been referenced for their work on 'planetary boundaries' reflecting an earlier concern expressed in the limits to growth by the Club of Rome in the 1970s and again in 2014. The planetary boundaries are intended to represent earth system processes, which, if crossed, could produce unacceptable environmental change potentially endangering human existence (Campbell et al., 2017). It is not surprising, therefore, that given the overbearing influence of biophysical perspectives that a linear view of integration as a panacea to the challenge posed by WEF-related planetary boundaries being reached has emerged.

The WEF Nexus school of thinking can be demarcated into two streams of thinking: (a) resource integration/linked cycles/resource interdependencies (Bleischwitz et al., 2018; Liu et al., 2018) and (b) governance of trade-offs, siloes in decision making and scope for synergies (Kurian and Ardakanian, 2015; Kurian and McCarney, 2010). Conceptually, the WEF resource interactions/integration school is different from the WEF Nexus in its emphasis on optimizing resource use as a response to deteriorating planetary conditions. The biggest contribution of the WEF Nexus governance scholarship has been to point out that without engaging with the political economy of the environment it would be next to impossible to ensure that resource integration alone will deliver sustainable development. This triggered a crucial turning point in moral realization as it has resulted in an urgent call to do away with continuing siloes in management of water, energy and food that were exacerbated by divergent interests. Policy makers and scientists see as inevitable that these three components would now need to be bridged in order to materialize sustainable development. In this regard, it is fair to argue that WEF Nexus governance scholarship has opened up promising new avenues by highlighting the role of social network analysis and agent-based modelling which our proposed framework has elevated by adopting the development of composite indices in monitoring social and institutional outcomes stemming from trade-off analysis. Along with this theoretical evolution, WEF nexus governance scholarship has put a spotlight on "food"- an output arising from a given mix of water and energy resources, inputs in the form of infrastructure construction, operation and maintenance with significant potential to produce distributional effects (through delivery of services for the poor) with serious consequences for political decision making. Two confounding questions that have emerged from the divergence in the WEF resource/integration and WEF Nexus governance scholarship are worth mentioning at this point: (1) why a focus on water, energy and food resources in particular? (Do et al., 2020) and (2) what about other factors such as land tenure, income and employment that have up to now received attention only as exogeneous variables? (Wilchens, 2017; Saidi and Elgalib, 2017). In a similar vein, others have pointed out that an important outcome of greater integration could be optimization of resource use as a panacea to the risks posed by planetary boundaries being reached. But others have bemoaned the fact that current models are unable to undertake longitudinal analysis of optimization pathways due to a lack of data (Liu et al., 2018).

Our own critical review of dominant models of WEF interactions (see Table 2.4) has led us to highlight confusion between the pursuit of integration in development/policy practice versus integration in science (via integrative modelling) (Kurian, 2017). Those leaning towards integrative modelling, however, too often mistakenly assume that more complete data and information would provide governments and decision-

Table 2.4 Example methods used in water, energy and food (WEF) nexus quantifications.

Method	Functioning	Examples
Biogeophysical model	Investigates biogeophysical processes related to FEW	Links hydrological (VMod), meteorological, floodplain (EIA 3D model) and climatological models (GCMs) to examine consequences of changes in a watershed on local economies (with respect to FEW resources). Hydrological models (HYMOD_DS) are used to simulate changes and implications for FEW in the Brahmaputra River Basin, South Asia under different scenarios.
Production model	Represents the amount of a FEW resource produced in different scenarios	Trade-off models that investigate rice paddy and hydroelectric production in Sri Lanka under different management regimes.
Life cycle assessment	Evaluates 'the inputs, outputs and potential environmental impacts of a product, process or system throughout its entire life'	Investigation of the impact of agricultural production on evolving renewable energy programmes on the FEW nexus in Qatar. Assessment of the impact of Kellogg Europe cereal production on FEW interactions (GaBi Software).

Continued

Table 2.4 Example methods used in water, energy and food (WEF) nexus quantifications.—*cont'd*

Method	Functioning	Examples
Ecological footprint (or water/ energy footprint) analysis	Evaluates the total environmental impact of a product or activity on FEW systems (normally in terms of area or natural capital)	Calculations of water and energy footprints for different agricultural products grown in Nepal. Estimates the water footprint of energy use in California.
Material or resource flow analysis	Quantifies flows and stocks or materials/resources in an FEW system	Analysis of water fluxes, deforestation and energy flows for cooking and heating in Uganda to discuss implications for food security.
Econometric model	Probabilistic modelling used to predict economic variables affecting FEW; often used for forecasting	Analysis of the impact of energy demand and water availability on food security in BRICS countries (Brazil, Russia, India, China, South Africa) using panel econometric models.
Cost-benefit analysis	Evaluates strengths and weaknesses of alternatives of a measure or action for FEW using a business framework	Evaluation of costs and benefits of alternative irrigation technologies for FEW in Nepal.
Supply chain analysis	Investigates inputs and outputs across all stages of a product's production as it moves from primary production through supplier to customer in FEW systems	Investigation of sources of waste in all three sectors of global FEW and targeting points in the supply chain to improve efficiency.
Input-output model	Quantifies the economic relationship (monetary flows) between two entities (or sectors) as related to FEW	Quantification of two-way interdependencies among food, energy and water and their implications for resilience of FEW systems; application to evaluating new policies such as organic farming.
Computable general equilibrium model	Estimate how an economy responds to changes in FEW policy or other factors by following a general equilibrium paradigm	Prediction of potential future scenarios of Australia's environment and economy under different FEW and climate conditions.
Agent-based model	Models actions and interactions among individual actors and their impacts on FEW systems	Analysis of diverse FEW-related factors affecting individual farmer decision-making in the midwestern United States with particular focus on biofuel crops.

Table 2.4 Example methods used in water, energy and food (WEF) nexus quantifications.—*cont'd*

Method	Functioning	Examples
Systems model	Examines relationships among multi-sectoral FEW systems, often incorporates scenario analysis and decision support tools	Multisectoral analysis of urban FEW use, flows and resource metabolism in London, towards strategies for better resource efficiency. Integrated assessment models such as PRIMA or CLEWs models that integrate across multiple systems (for example, climate, hydrology, agriculture, land use, socioeconomics and energy systems) to make policy assessments. Other systems models include BRAHEMO, WEF Nexus Tool 2.0.

Many studies integrate multiple different methods (for example, physical models and economic models were integrated to create multisector systems models).
Source: Modified from Liu et al. (2018):470.

makers with incentives to properly target financial expenditure for environmental management (Keairns et al., 2016). Such assumptions however, do not reveal the recursive and non-monotone dimensions of environmental decision-making processes that we pointed out earlier for two reason; 1) framing-problem- type of data and information that the conventional school of modelling has been utilizing is not relevant for providing solution for emerging shortcomings of current environmental management; 2) dynamic features of political economy of public decision making represented in the forms of electoral systems, mechanism of budget appropriation and individual discretion which require much nuanced interpretations by cultural considerations as those that are driven purely by considerations of economic rationality -optimization and efficiency gains (Kurian, 2020; Weitz et al., 2017) and thus necessitates a paradigmatically different approach from those leading the conventional modelling. This is precisely why the WEF interactions framework falls short in capturing the fluid, adhoc and non-sequential characteristics of processes entailing environmental governance. Furthermore, emerging policy challenges in environmental policy suggest that there is a greater need for more substantive theorization in the area of political economy of public decision making in environmental governance by the WEF Nexus governance scholarship which we intend to contribute through this book.

Against this background, we define the nexus as a framework for integrative modelling of trade-offs with the objective of advancing synergies in decision making on "water-energy-food interactions"… (Kurian et al., 2019). While the integrative modelling is centred around the political economy considerations in our proposed nexus framework, here we turn to the body of knowledge constituted from three schools of propositions on environmental governance in which multi-faceted features of political economy of environmental decision making have evolved around theoretical discussions in the area of: property rights (CPR), behaviour of administrative agency and consumers/service users (ABM) and the role of data-sharing for the coordinated decision-making processes (SNA). Here, we are aware that the knowledge emerging from these three lines of theorizing are drawn exclusively from community-based perspectives, missing the connection to higher order institutional rules. In contrast, scholarship on policy science is focused largely on higher order rules of public finance. Therefore, we strive to combine the insights of these binary perspectives to offer a more holistic understanding of political economy considerations in environmental governance. Let us start with the CPR in the following section.

Common property resources and environmental conservation – Gaps between theory and practice

CPRs point to biophysical resources that are characterized by non-excludability and subtractibility in their use. In other words, it is difficult to exclude individuals from using them, and every unit of a resource that is consumed by person X means there is less of it available for others to use (Ostrom, 1990). When institutional arrangements for management of CPRs are weak the world's forests, lakes, groundwater aquifers and grazing pasture will become open-access resources and pose serious problems to sustainability of eco-system services and livelihood systems (Ostrom, 1990). Under the CPR body of knowledge, the issue of property rights and land tenure weighs significantly. It highlights the fact that different structures/systems for executing property rights within states can mediate the relationship between public policy and environmental outcomes. For instance, communist states were known to own all land and discourage private property rights. Several other States such as India have maintained ownership rights while establishing usufruct rights that allow communities to share in the use of the products from publicly owned land – for example, rights to fuelwood or timber (Sivaramakrishnan, 1995). Further, Esther Boserup in her seminal work on demography and the environment alludes to the role of share cropping systems for both private and communal land especially in the humid subtropics to highlight the role of land tenure systems in management of soil fertility (Boserup, 1990). Others have examined how inter-generational change in land tenure systems can affect land productivity, wealth inequality and distribution of poverty among women headed households (Bardhan, 2005; Zwareveen and Meinzen-Dick, 2001).

The issue of property rights is especially important when the resource under consideration is exhaustible but renewable (Baland and Platteau, 1996). For example, when fisheries are overharvested and soils are overused, conservation is key to ensuring that one can find a pattern of positive use such that its stock does not shrink over time. Every ecosystem has its own *carrying capacity*, a measure of the amount of the natural resource that can be exploited without endangering the reproduction (renewal) of the system. But beyond the issue of carrying capacity, it is important to understand that not all resources can be used as a productive unit – some resources such as clean air and water need to be conserved because they safeguard health and provide recreational services. Furthermore, contrary to many environmental models, one cannot always substitute capital with environmental resources in the production functions (Barbier, 1997; Dasgupta, 2004). Finally, since changes brought about by an optimal exploitation of environmental resources are slowly cumulative, there may be 'threshold levels of exploitation', for natural resources such as land, forests, pure air or clean water that experience ecological processes (examples include precipitation and evaporation) beyond which environmental threats such as droughts can become magnified.

There has been a rich body of theorizing on the issue of property rights and environmental conservation that has debated the comparative strengths of public, private and communal property rights on conservation of renewable resources that are prone to exploitation over time. But although appealing, several of the arguments betray a gap between theory and practice; for example, successful cases of community-led forest management remain disparate and isolated because the normative framework in many instances has failed to gain community support for large-scale implementation based on widespread adoption of the principle of subsidiarity (Gibson et al., 2000). Infrastructure construction and maintenance as it relates to the management of wastewater offers another example of how a supportive normative framework can advance sustainability. In institutional terms, the opportunity for translating the potential for wastewater reuse into practice in the Netherlands, for example, to address challenges of water quality decline and scarcity hinges upon the following considerations: (1) norms that would facilitate integration of water resources management from source to reuse-addressing issues of sectoral water allocation; (2) norms for costing of water supply and sanitation interventions that would reflect costs of separating waste at source and (3) norms for billing of water supply and sanitation services, especially in contexts where multiple service providers from public or private sectors are involved (Salome 2010).

Carrying on in a similar vein, norms reflected in global monitoring regimes such as those that relate SDG[5] 6.3 do not acknowledge important political economy considerations as is evident from the following three shortcomings. First, the target for SDG 6.3 that focusses on the goal of wastewater reuse is focussed only on biophysical aspects of wastewater use (Kurian et al., 2019). Second, the indicators do not explicitly consider the issue of wastewater reuse- does domestic wastewater exhibit characteristics of an open-access resource and if so what incentives are required to reverse the negative externalities with implications for public health and the wider economy? Third, the monitoring methodology is biased toward reporting status on wastewater use and not toward understanding 'the incentives' that would facilitate wastewater reuse. In this connection, we argue that the effectiveness of SDG 6.3

[5] The SDGs were agreed by UN member states at the High-Level Political Forum (HLPF) in September 2015. SDG target 6.3 states 'by 2030 improve water quality by reducing pollution, eliminating dumping and minimizing release of hazardous chemicals and materials, halving the proportion of untreated wastewater and substantially increasing recycling and safe reuse globally' (UN-Water, 2015). The SDG target 6.3 by methodologically implying 'wastewater supplied to a user for further use with or without treatment and excludes water which is recycled within industrial sites' hints at the potential for wastewater reuse in agriculture (WHO UNICEF, 2015).

as a leading global normative framework fails to acknowledge the trade-offs involving biophysical and administrative scale, between local and global levels and betrays a sharp divergence in the interests of key institutions, i.e., global scenarios emphasizing norms of resource reuse, recycle, retrofit, remanufacture and reduce versus regional and local scenarios emphasizing service delivery norms that reflect a concern for affordability, coverage and reliability of public services.

The divergence between global and regional scenarios is bound to lead to unintended consequences and rebound effects of policy intervention. It is in this context that public choice theorists have debated the merits of decentralized monitoring of common property resources such as fisheries, underground aquifers, livestock pastures, air quality, forests and irrigation systems (Bromley, 1992). Here, it has been pointed out that resources such as lakes are more easily amenable to decentralized monitoring as against underground aquifers since extraction of resource units is easily visible and its use is localized. 'In the case of surface irrigation systems, for example, pumping of water is relatively easily observable as each irrigator keeps an eye on his neighbour's actions. When water rotation is organized, monitoring is costless: the presence of the first irrigator deters the second from an early start, the presence of the second irrigator deters the first from a late ending' (Baland and Platteau, 1996:315–16).

Others have argued that decentralized monitoring may not be reliable, especially in the case of repeated interactions. Apart from the fact the violations of resource extraction rules may not be reported at all, in other cases, violations may not reported owing to the fact that it may incur the wrath of the person whose behaviour is being monitored. It has also been pointed out that larger groups would entail formal guards to monitor and enforce rules while relatively smaller groups can ensure monitoring without recourse to formal guards. Elinor Ostrom offered optimism about the prospects for decentralized monitoring when she outlined eight design principles of successful decentralized monitoring. One of the important design principles related to the fact that monitoring is successful when smaller groups are nested within larger systems for appropriation, provision, enforcement and conflict resolution (Ostrom, 1990:90). Other design principles deemed important for decentralized monitoring to succeed are threefold. These include; 1) the existence of clear boundaries for common property systems; 2) minimal recognition of rights of appropriators to devise their own rules which are not challenged by external authorities and 3) violators of rules are susceptible to graduated sanctions.

Apart from the issue of behaviour change it is imperative that we examine other opportunities and pitfalls that we have to take note of when we engage in instituting policy change. From an analysis of contrary cases where no correlations exist between poverty and the environment it can be concluded that although the poor's lack of resouces to engage in environmentally sustainable activities (Jaganathan, 1989; Jodha, 1990) could be a fact, research that attempts to better understand causality can be a useful first step towards designing robust monitoring and feeback systems.

In this connection, a propositon provided by Dasgupta et al. (2005) may be resourceful here. They suggest that at least three inferences can be drawn in the senario where you see negative co-relation between the poverty and the environmental degradataion: (1) the country's official statistics may not provide a dis-aggregated view of the interactions between poverty and the environment, (2) the government may have already addressed them effectively but does not publish information that would make it possible to ascertain what steps were taken and the impact of the intervention over time and (3) the nexus may be operative only partially at the district or subregional level and mechanisms do not exist for other districts/regions to cooperate to ascertain poverty-environment causality based on shared interests. We are optimistic that in the absence of clearly verifiable data, tapping into epistemological knowledge by adopting a panel composed of local experts could clarify lines of causality at various spatial levels of planning and management. This could be one way in which scientific inquiry can contribute towards policy change through support for 'adjusting the value system to the reality system and how when viewed in this light it becomes more relevant to talk not of goals and objectives that are achieved, once and for all, but of norms and standards that are maintained or modified over time…as a result, the essence of policy-making becomes a sequential process in which public policy is being formed as it is executed and executed as it is formed' (Gregory (1997): 188). Therefore, the science that guides policy-making, monitoring and evaluation cannot be one that is focussed merely on the testing of scientific theories but on drawing lessons from the design of more topical, effective and cost-effective interventions and policy instruments that address challenges along the poverty–environment continuum. This can contribute to a theory of change on the environment-development Nexus.

Agent-based modelling of water–energy–food interactions: Meta-data, integration and synthesis

To begin, what is agent? Often in the discussion of environmental science, there is a confusion between the definition of agents and stakeholders, using these two terms interchangeably. Here, agents/agencies are defined as public organizations such as the government and public goods research bodies. Stakeholders on the other hand include farmers, community leaders, etc. who potentially benefit from policy and technical interventions. But there are important differences between the behaviour of stakeholders and agents. Agents operate within the framework of bureaucratic rules: reporting protocols, performance standards, clearly defined roles and norms for promotion. This is in contrast to stakeholders who besides having to conform to community norms are free to behave according to their free will while agents conform to the rules of agencies (Fukuyama, 2016). Within agent-based models, agents are defined as autonomous decision-making algorithms. By focusing on interactions between agents who are boundedly rational and vary in their attributes within the agent population, agent-based modeling has the potential to generate a series of observed behavioural regularities that may be useful in clarifying the following issues: (a) how do agents make decisions? (b) how do they forecast future developments? (c) how do they remember the past? (d) what do they believe or ignore? (e) how do they exchange information? (Poteete et al. (2010), p. 211).

Agent-based models rely on meta-analysis of N-cases to deduce patterns of agents behaviour. But the non-monotone and non-recursive nature of human-environment interactions, especially when the past may no longer be a reliable guide to the future makes it imperative that we proceed with caution with the design of agent-based models. To better understand this challenge for modellers who are keen to inform environmental and governance through their work, it is useful to undertake a short excursion of the divergent paths that conventional approaches to institutional analysis have taken. The theory of public goods has highlighted the fact that certain resources by being non-rivalrous, non-excludable and subtractable are potentially susceptible to exploitation (see Fig. 2.1). Examples of such common property resources include flood protection, clean air and water quality protection, forest, aquifer and rangeland protection. Governments by being able to take on risks that the private sector or individuals would not be able to take on are therefore responsible for management of externalities and transaction costs associated with bureaucratic design and enforcement of regulatory instruments such as standards, guidelines, notifications, directives and circulars (Stiglitz, 2000).

The task of monitoring externalities is, however, complicated because of the nature of goods and associated interactions. Some goods such as groundwater aquifers are fungible resources compared with lakes and forests. This is because as we pointed out earlier it is relatively easy to monitor use of the resources that are visible when compared with fungible resources where it is difficult to monitor when resources are being extracted, by whom and to what extent. Therefore, public choice theory has been successful in employing case studies of successful environmental management to forcefully argue for a role for community participation in stewardship of non-rival public goods. Nevertheless, the success and failure of communities to engage in cooperation can become more complicated by the repeated nature of interactions. Here, public choice theory has argued that provision of people with vital information relating to the characteristics, location and available extraction units will lead to rational strategies for co-management of environmental resources. Entitlement and New Institutional Economics (NIE) scholars have been able to demonstrate, however, that not all cases of successful co-management can be explained by rational action based on recourse to complete information (Mosse, 1997; Bates, 1995).

The divergence in approaches to study of environmental decision-making has implications for methodology. Game theory has relied on deductive reasoning to tease out generalizable principles and test hypothesis based on postulates of rational action for repeated interactions. ABM has carried on in this vein by working with metadata to develop and test assumptions regarding environmental decision-making at the level of both public agencies and community groups (Potette et al., 2010). But the assumption that people will act rationally in situations of incomplete information is only one of the pitfalls of a game-theoretic approach to modelling cooperative outcomes. Questions such as how do agents make decisions? (b) how do they forecast future developments? (c) how do they remember the past? and (d) what do they believe or ignore? are all questions that point to the significance of understanding the role of hierarchy, history and official discretion in shaping institutional behaviour. Entitlement scholars should therefore be credited with relying on variants of longitudinal case studies to understand the role of power, social exchange and historical relations among political factions in explaining success and failure of co-management strategies (Bebbington, 1997; Leach and Mearns, 1996). When examined within a framework of repeated interactions, analysis of power and social exchange is more likely to highlight the importance of factors such as a history of past cooperation, trust, reciprocity and reputation and social standing in mediating successful cooperative outcomes.

One of the abiding points of contention between both these groups of scholars was on the relative merits of focussing on efficiency vs equity concerns of environmental management. This tension was best captured by the hypothesis about the effects of group heterogeneity on collective action (Poteete and Ostrom, 2004). The hypothesis captured the underlying analytical differences in the examination of issues of structure and hierarchy involving both agencies and groups. (Kurian and Dietz, 2012). The hypothesis demonstrates the multi-dimensional aspects of agency behaviour by emphasizing the importance of aligning rules within nested organizational structures (Ostrom, 1990; Gibson et al., 2000). In Chapter 3 we will elaborate on this issue by delineating the causes of institutional failure through analysis of 5 typologies. The analysis will show that ABM is therefore confronted with an important challenge with potential to enrich linkages between assessment methodologies and policy science in general and the theory of public choice as it relates to management of common property resources in particular.

The challenge can be explained as follows: should modelling draw upon metadata to deduce the behaviour of agents under strict experimental controls or should theory develop based on a meta-analysis of longitudinal case studies that serves to pilot-test institutional instruments such as directives, notifications, guidelines, circulars and standards? (Albrecht et al., 2018). There have for instance been case studies that have examined how stall feeding of cattle can impact upon the condition of sapling regeneration in forests (Kurian and Dietz, 2012), while some have also examined their implications for arriving at similar conclusions via large N-studies with the propensity to be easily upscaled via targeted policy interventions (Poteete et al., 2010). Our optimism in the latter strategy stems from its promise of synthesizing data and knowledge to support the identification of an incentive structure whereby robust alignment of rules between groups and regulatory agencies can flourish. Further, from the perspective of designing effective policy instruments, longitudinal case studies may be better placed in understanding the predominance of factors such as factional conflict and past history of cooperation in predicting potential for collective action in the future. This trend is further substantiated by the emergence of futures studies and machine learning methods that are better equipped to incorporate systemic complexity and interactions among mutliple datasets and models (Scoblic and Tetlock, 2020).

One issue which is not sufficiently addressed by the school of agent- based modelling is the relative characteristic of public policy choice which is highly value-laden (Kuhn and Wolpe, 1978). The Bottom line is that we all wish to mitigate environmental trade-offs by seeking to avoid the most extreme forms. Nevertheless, what is considered less extreme form of trade-offs is a highly relative issue while there is no one-size fit-all solution when it comes to mitigating institutional and environmental trade-offs since local specificity (combination of geopolitical and socio-ecological factors) determines the very nature of such outcomes. Public policy choice dynamics are thus shaped in the struggle of what should be done and what can be actually possible to do. Agent modelling which seeks a solution in the area of public policy choice should reflect this point and seek a way to frame this dilemma in a more nuanced way.

Social networks, rationality and bounded interactions: Structure, hierarchy and change

SNA, which has approached institutional analysis by emphasizing the structural nature of a given social network, is an example of an instrumentalist view of the uses of data and method in environmental policy and governance. SNA has demonstrated the extent to which an overall network is densely or weakly connected (Berardo and Lubell, 2016). It can also identify the most centrally connected nodes or nodes that are connected to these central nodes. SNA data can be used to create 'maps' that depict types or strength of relationships and transactions between and among people and organizations.

The instrumentalist logic of SNA can lead one to conclude that where social networks are weak or interactions less frequent, coordination mechanisms can be stipulated by external agents to draw up regulations and build institutional capacity to ensure sharing of information among public agencies and between public agencies and consumers (Waylen et al., 2019). What such analysis misses however, is related to social location and associated power of the agent in the context of a hierachical organization that is shaped by the administrative structures and processes. The power of agents to actively participate in the design, monitoring and reform of institutional procedures is shaped by where they are located in the hierarchy: local versus state government, finance ministry versus agriculture ministry or landed or landless farmer. It is no surprise therefore that Bina Agarwal (2000) and others have forcefully argued for the need to examine the gendered impacts of development projects to better understand the role of hierarchy in development planning.

In a similar tone, density of ties does not simply imply that coopeartion within the network will emerge because of the nature of aligned resources and interests of members of the network to achieve cooperation, which is our second point. In this context, one of the strands of SNA that has played an influential role in developing models of environmental decision-making is the one that is focussed on distribution of resources and interests (Oliver and Marwell, 1993). Theorizing on *conjunction of probabilities* essentially argues that for collective action to emerge, it is not sufficient only for individuals to be well endowed. For well-endowed individuals (those with power relative than others in public agencies or community groups) to succeed in mobilizing others, they must also have sufficiently high levels of interest, leading to a conjunction of probabilities that will succeed in achieving a cooperative goal. The distribution of resources and interests within groups can shape trade-offs: whether to prioritize the group versus individual self-interest. Sometimes a conjunction of probabilities can give rise to incentives for leaders to take the initiative to organize the resources (labour and time) necessary to provide a service for the larger community, thereby mitigating the effects of trade-offs in environmental management with overall benefits for the group (Kurian et al., 2018).

The WEF Nexus: Framework integrating agency behaviour analysis for studying trade-offs in environmental governance

The analysis above emphasized four concepts: social network analysis, common property resources, agent-based modelling and Water-Energy-Food (WEF) interactions/resources. Each of the above concepts has the potential to further bolster the effectiveness of WEF Nexus theorizing. Rigorous application of SNA can help us understand prospects for data sharing with the goal of promoting coordinated decision making. Common Property Resource literature makes us realize that in the same way that as resources and interests can define trade-offs within communities, similar considerations can arise in the context of public agencies. When different agencies make appropriation requests for annual budgets, they are in a sense competitors for scarce resources. Although different public agencies are similar in terms of bureaucratic rules, the fact that they are competitors for a scarce resource makes it difficult for them to cooperate by way of contributing jointly towards achieving a common policy goal. This is because cooperation may necessitate savings of budgetary resources which go against the logic of annual appropriations – which is to always ask for more resources and not risk shrinkage of the department itself (Bjorkman, 2010). This competitive environment makes it difficult for trade-offs to be mitigated but instead for unintended consequences of departmental action to be exacerbated.

Our previous research on environmental policy and governance has highlighted the importance of administrative boundaries and issues of hierarchy, structure and change. Far from viewing environmental decision-making as being constrained singularly by data or capacity, the nexus approach by placing institutional trade-offs at the centre of agent-based modelling exercises does not underplay the importance of data or improved technical capacity (Twisa, 2021). Instead, it offers enormous opportunities for policy-relevant research through pilot-testing specific institutional instruments with the potential to change individual and agency

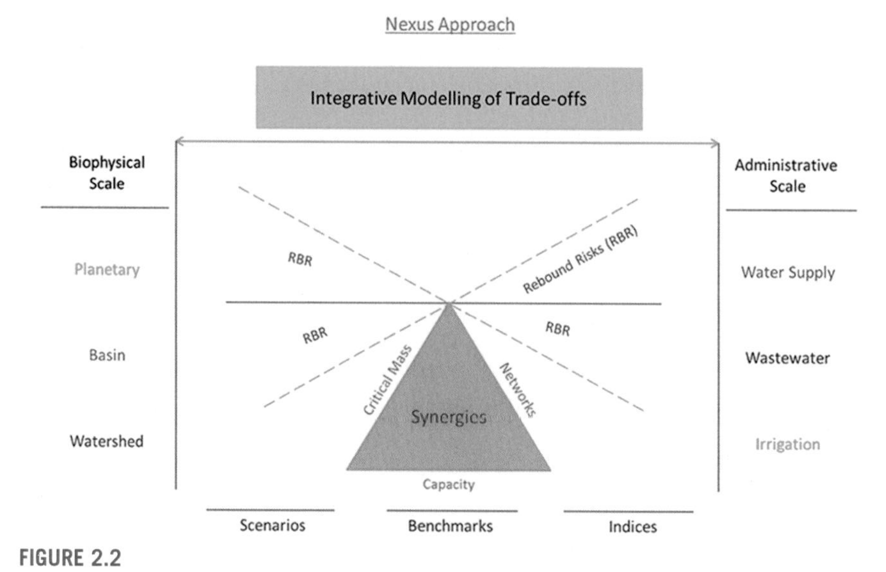

FIGURE 2.2

Nexus framework.

Source: Kurian et al. (2019):5.

behaviour as they relate to environmental planning and management (see Fig. 2.2). This is because the Nexus by emphasizing integrative modelling of trade-offs with the objective of advancing synergies in decisions on WEF interactions offers several opportunities to address the tensions between equity and efficiency considerations in the design of environmental models that can inform decision making (Kurian et al., 2019). These challenges include the inability of uni-dimensional environmental models to deal with the uncertainty posed by exogeneous variables on account of weak feedback loops between developmental interventions and their effects on a heterogeneous socio-ecological landscape (Yang et al., 2016).

From a policy perspective therefore, the Nexus framework is more likely to advance the use of concepts such as thresholds and critical mass in examining the role of financing, technology adoption and path dependences of institutions in sustainable development (Vogeler et al., 2019). Such an approach has the potential to overcome a serious shortcoming of game theoretic approaches that assumes that all participants have equal endowments and interests in pursuing a cooperative goal (Dick et al., 2018). Furthermore, most impact assessments of technical interventions in agriculture, for example, have found that the inference problem is widespread because most research for development assumes that technical options such as improved crop varieties, livestock breeds, agronomic practices and management models would have the effect of mitigating levels of poverty (Tomich et al., 2018).

Nexus analysis by broadening the focus of longitudinal assessments to enable comparisons of organizational performance that enable the identification of incentive mechanisms, however, can help understand the non-linear, non-recursive and non-monotone effects of environmental policy and governance. For example, it is critical to understand that the competitive budgetary strategies of public agencies can stand in sharp contrast of Non-Governmental Organization (NGO) initiatives that target equity concerns of project beneficiaries (Meinke et al., 2006). Furthermore, of particular importance is the role that the nexus approach can play in developing methodologies for co-curation of data and downscaling of environmental models with the potential to effectively monitor and evaluate the impact of interventions that pursue sustainable development (Agrawal, 2020; Gebrechorkos et al., 2019). This is where the nexus approach by enhancing the policy relevance of research by forging robust linkages between biophysical and institutional assessments can thereby improve the salience, credibility and legitimacy of public goods research.

Monitoring and evaluation of development interventions is premised on the idea of monitoring the behaviour of agencies. But it is important to recognize here that the tension between efficiency and equity may be exacerbated by the self-preservation logic of the bureaucratic apparatus which is framed by the norms of bureaucratic conduct: selection of agents based on merit, promotion, decision-making based on consensus, decisions guided by policy guidelines and operational manuals, clearly defined penalties for non-conformance with personnel management rules and information gathering to account for revenue and expenditure appropriated via public budgets (Fukuyama, 2016). The most pertinent items to monitor in this context therefore, is the quantum of appropriations via public budgets, recurring expenditure items, commitments to operation and maintenance of established infrastructure, commitments to staff training, organizational organigrams, levels of staff discretion with regards to implementation of policy directives and extent to which spending is off-line from appropriated budgets. The extent to which bureaucratic apparatus responds to the such "soft" concerns would serve as commentary on the extent to which the trade-offs between efficiency and equity are being addressed. When infrastructure projects have elements that involve participation of private sector and community groups, then monitoring and evaluation would need to pay attention to citizen perspectives on coverage, affordability, quality and reliability of public services. For this purpose monitoring regimes would need to translate specific activities, outputs, outcomes and impacts specified in public-private partnership contracts and vendor service agreements into metrics that can be justified for their precision in tracking the cost-effectivess and distributional impact of development interventions focussed on advancing sustainability (World Bank, 2004, 2009).

5. Organizations: Incrementalism, boundary spanning roles and feedback loops

So far, we have reviewed theoretical trajectories related to the concepts of norms and institutions. To recap in simple terms, norms are shared values while institutions are a set of rules. In this connection, organizations, which compose the last pillar of three key concepts of boundary science, are defined as the agency – public organizations such as the government ministries and bodies and boundary organizations (or think tanks as we indicated in Chapter 1), that typically engage in the public goods research, i.e., CGIAR and even NGOs, which are mandated by a particular government ministry to undertake that role. These organizations are assigned to implement rules (institutions), which promote policy outcomes in the context of environmental sustainability. To this end, our discussion in this section will focus on the role of organizations from the perspective of emerging principles which should guide their evolution and fundamentally change the way they engage in the business of decision-making as it pertains to environmental sustainability and sustainability research.

Incrementalism: Implications for modelling of environmental decision making

Our previous research experience leads us to conclude that conventional approaches to environmental modelling are unidimensional because they emphasize: (1) a disproportionate focus on analysis of behaviour of biophysical resources; (2) efficiency of ecological systems only; (3) potray a bias towards use of statistical analysis to examine interactions between SDG goals and targets and (4) case study research data, models and approaches that have neither been pilot-tested nor valorized through engagement with governance structures and processes (Cai et al., 2017; Bleischwitz et al., 2018; Dombrowsky and Hesengerth, 2018; Liu et al., 2018; Scott et al., 2018). Thus, conventional approaches to environmental modelling would promote siloes in environmental planning and management with potential to seriously undermine the credibility of the global monitoring regime.

In this connection, multidimensional modelling approaches can draw upon several insights offered by theorizing on 'incrementalism' to highlight the conditions for success and failure in environmental governance (Gregory, 1997). Incrementalism essentially argues that environmental decision-making is seldom linear, monotone and non-recursive because of the continuous tension in environmental decision making; choosing between technical rationality (i.e., selecting the appropriate means for achieving given ends) and economic rationality (i.e., determining the most efficient use of resources among competing purposes). The key elements of incrementalism that are worth emphasizing at this stage are characterized as status quo, marginal trial, ad hoc and politically driven. Details are described as follows:

- Status quo: That policy-makers typically confine themselves to consideration of those variables, values and possible consequences that are of immediate concern to themselves/their agencies and which differ only marginally from the status quo, thus greatly simplifying their analysis of possible options.
- Marginal trial: In the face of limited information and theoretical input, policy movements are based on trials and error interventions of an intendedly marginal kind, so that unintended consequences may be coped with more easily.
- Ad-hoc decision: That political and policy change is not a function of any coherent set of transcendently guiding principles/goals.

- Political more than practical: That policy-making is a process of political and social interaction – negotiation and, bargaining among groups with competing interests, differing values and performing different roles in a hierarchical power structure.

Mindful of its conceptual limitation that failed to include complex power interplays outisde of the technical-economic environmental decision-making framework (i.e. corruption based on personal or group interests and favours), incrementalism can be useful to illustrate the divergence between models that are guided by theoretically informed scientific analysis and those that are driven by what is politically expedient and/or practically possible. While political economic issues constituting the environmental decision-making climate will be addressed in the later discussion, the applications of incrementalism in environmental modelling cannot be underestimated especially in the context of recent analysis that has called into question the connection between environmental management and development. The 'Poverty–environment' nexus is an appropriate example to be elaborated in this context. There has been an implicit assumption that there is a link between improved management of for example, air quality, water quality and soils and impact on poverty. A critical review of data sets for Cambodia, Vietnam and Laos by (Dasgupta et al., 2005), however, has called into question this assumption. With the exception of water and sanitation in Vietnam where greater state intervention in the sector explained positive impacts on levels of poverty, there did not seem to be any correlation between concentrations of poverty and air quality and soil erosion in the region. The authors therefore suggest that given this poor correlation between poverty and environmental management, public budgets should be split between the departments and activities focussed on both dimensions. What is more striking here is that they pointed out that the poverty–environment nexus is important only if it has consequences for the allocation and administration of public resources for alleviation of poverty and environmental problems (ibid:619). This conclusion dovetails with the suite of prescriptions that have been tested by international organizations with reference to specific institutional arrangements relating to budget support, output-based aid and Cash Conditional Transfers (Iyer et al., 2005; Agrawal, 2020; ADB and WSP, 2010; WSP, 2008; WSP & PPAIF, 2009; see Fig. 2.3). The example of OBA in Fig. 2.3 emphasizes the importance of dovetailing external aid with public budgets

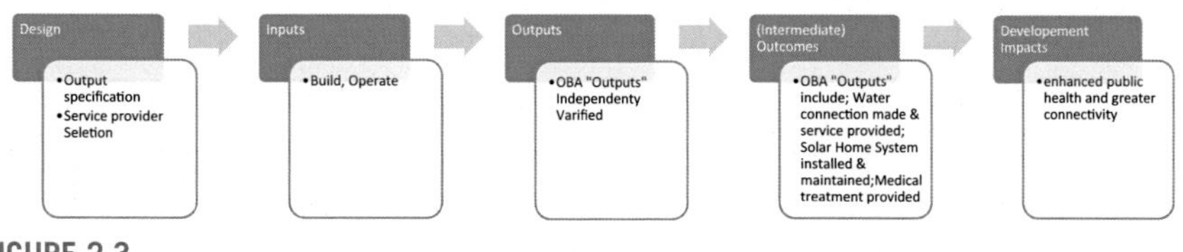

FIGURE 2.3

OBA contracting spectrum. *OBA*, output-based aid.

Source: The World Bank (2009):19.

so as to enhance the chances of policy impact that broadly align with existing societal norms and institutional structures.

Boundary spanning roles in boundary organizations – A two-way dialogue

There is growing convergence of trends based on empirical work on the nexus and theorizing on common property resource management (Kurian and Dietz, 2012). On one hand, there is a growing consensus that environmental models that specifically address societal or public policy questions have greater potential to enable decision-makers to prepare for and effectively mitigate trade-offs in environmental decision making. Mitigating trade-offs entails identifying taxonomies of interdependencies between water and energy that are not limited to analysis of resource interactions. For this, there exists no ideal scale at which a problem can be addressed (Urbinatti et al., 2020).

On the other hand, a post-Ostrom research agenda that has begun to take shape emphasizes that rather than adding more variables, progress requires a clearer, more consistent approach to selecting, defining and measuring institutional elements; stronger links between theory and empirical research; a greater focus on mechanisms and causality and the development and application of new methods, including quantitative and mixed method approaches to studying institutional arrangements for management of common property resources. Strengthening the connections between theory, models and data suggests several promising avenues for advancing institutional analysis through the study of relationships between institutional structure, process, function, context and outcomes (Cumming et al., 2020). Typology construction followed by composite indices supported by open access platforms fitted with open-source modelling interfaces offer an opportunity to build an inclusive, cost-effective and transdisciplinary research process focussed on addressing societal challenges.

Strengthening the connections between theory, models and monitoring approaches pre-supposes the selection of individuals with boundary spanning skills in boundary organizations. These skills could include an ability to scan the horizon to identify projects where it is possible to link research questions to actual policy concerns. This may mean starting with small scale and clearly defined questions derived from scientific discussions and expanding out by linking to specific policy concerns at a larger scale – either regionally or globally. This would entail that individuals have the ability to seek political buy-in at early stages for the results of the research, be able train practitioners on specific topics such as design of a Delphi assessment and to seek to engage citizens and beneficiaries of development programs in the design, implementation and evaluation of interventions. Boundary spanning skills would also include the ability to communicate the impact of the project to a non-scientific audience by leveraging data visualization, artificial intelligence and remote sensing tools. These are issues that conventional students of environmental modelling would have to consider as they prepare to build multi-dimensional models that can inform and be informed by policy processes at different scales.

There is an underlying question that has accompanied previous approaches to environmental modelling: what should be the goal – forecasting future demand or supporting long-term decision-making regarding infrastructure investments[6] within the framework of the environment–poverty nexus? In this book, modelling is framed as ways of re-imagining (not necessarily forecasting) future demand. We emphasize an approach advanced by Sharmina et al. (2019) whereby modelling is viewed as a mechanism that enables decision-makers to pursue transitional planning. While being quantitative, through inclusion of machine learning techniques it is possible to also incorporate qualitative or mixed methods with the goal of not providing 'one' answer in the tradition of one-size-fits-all and best practice exchange but a range of plausible representations of technical and financing options that can effectively mitigate trade-offs. Such an approach to modelling fits well with the requirements of decision-makers which is for the supply of a range of possible options from which to choose from by considering both what is scientifically advisable but also politically expedient. But this does not mean that either one or all of those options will be adopted by decision-makers based on the relevance or precision of the research undertaking (Mohammadpour et al., 2019).

Historically, environmental models have not been adept at accommodating for exogeneous variables and for rapid changes in the institutional environment. This is primarily because modelling exercises have been averse to the effects of uncertainty: parametric (uncertainty in a model's parameters, on inputs such as weather forecasts or population projections that have a range of potential future values rather than a single value) and structural (uncertainty about whether a model appropriately reflects the real world) covering, for example, a range of scenarios including peak energy consumption, informal economies and inequality, penetration of renewables and non-point sources of water pollution (Sharmina et al., 2019:20).

It is in this context that some have emphasized that social learning entailing a 'two-way dialogue' running from society at large towards researchers and policymakers (about norms and framings that should be emphasized in the production of science) and from experts to society about the detailed contents of the research that emerges which can potentially play a role in embracing exogeneity in model calibration (Urbinatti et al., 2020:7). Such a dialogue is bound to highlight the challenges relating to transdisciplinary scholarship covering issues such as language, skills and the nature of evidence in policy-oriented research (Harriss and Lyon, 2014).

[6] For an interesting discussion on climate-induced stranded assets (See Muldoon-Smith and Greenhalgh, 2019).

Learning about feedback loops in the environment–development Nexus

We pointed out earlier that policy is a normative expression of the will of government. Conventional approaches to uni-dimensional environmental modelling cannot be farther away in terms of contributing to pressing policy challenges such as climate change and poverty. To this end, there are three considerations that can contribute towards enhancing the relevance of environmental modelling to addressing pressing global challenges. They are (1) downscaling global models to better align with regional perspectives and policy challenges, (2) coupling biophysical and institutional models to be able to better prepare and respond to environmental risks such as droughts and floods and (3) strengthening feedback loops between policy interventions and environmental and socio-economic outcomes. Needless to say, these considerations necessitate significant methodological shifts in the way we engage in research, and we will spare this issue our attention for until later on in this section. But at this point it would suffice to state that boundary organizations will have to engage in definitive changes in modalities relating to skills training, partnerships and recruitment spanning different disciplines, geographical regions and institutional settings to be better prepared to address the challenges at hand.

Nexus studies have previously claimed that interdependencies or feedbacks are important and originate in the water and energy sectors (Bazilian et al., 2011; Marsh, 2008). But a thought-provoking comparative assessment of nexus feedback loops in China and the United States points out that the fundamental conceptual flaw in the current WEF interactions framework is that 'public policies aimed at enhancing virtuous cycles or avoiding vicious cycles tend to focus on water and energy sectors instead of concentrating on manufacturing industries' (Vivanco et al., 2018: 10). This is because contrary to earlier assertions economic activity does not originate from direct feedbacks arising from resource use but rather from dependent relationships involving the manufacturing sector. Further, water and energy are related to resource use, while food is related to service, which is made available through processing and production. Thus, it is not logically sound to cluster these three concepts into one framework of analysis. The example demonstrates that in China although interdependencies between metal production and metal processing are the strongest, yet the most important feedback loops are between cultivation of paddy rice and pig farming. In this case, the authors point out that the cultivation of rough rice, which requires blue water, is included in pig diets, whereas pig manure, which requires primary energy to be produced (via heating, ventilation and feed production), is used as fertilizer in rice fields.

In this respect, there are two significant implications we can draw from the analysis on the potential role of boundary organizations in advancing innovations in tracing policy impact. They are (1) design of consumer-oriented policies that lead to optimization of trade-offs and co-benefits between water and energy-and (2) targeting of key sectors through which water and energy propagate to final products to encourage a shift in fertilizer types, crop species and primary energy mix (see also

Goldenberg et al., 1990). In a similar vein as (Sharmina et al., 2019), we argue that modelling exercises should demonstrate how WEF interactions vary across time and space to ensure that policy and scientific discussions shift the focus from the supply of resources to their effects. All these insights provide greater impetus for novel approaches to downscaling, coupling and strengthening feedback loops in environmental modelling. This will be the focus of the following discussion.

Feedback loops, modelling platforms and environmental back-casting

Modelling can support a two-way dialogue through a paradigm shift. Rather than attempting to *forecast* resource use and then plan accordingly, modelling should seek to assess and challenge policy and planning options in relation to pertinent parameters that cover a range of issues such as climate change impacts, capital and operational costs of infrastructure, legislative and sociocultural shifts (Sharmina et al., 2019:25). This would mean that the scientific enterprise would have to shift away from modelling of bio-physical resources that preserves a 'shared technical interest' of both policy and methods in producing incremental change (Mohammadpour et al., 2019). Part of this paradigm shift can be encapsulated in what has been called a 'co-evolution' principle to modelling that reflects the way that infrastructures, technologies, institutions and practices jointly develop in a non-linear manner over time. In Chapter 3, we analyze five cases to show how a critical mass of financing and technology, thresholds to public action and siloes can combine depending on institutional trajectories to contribute towards synergies in environmental planning and management.

The concept of co-evolution seeks to capture key interactions, relationships and feedback loops between variables specified within three attributes of models: stochastic events, diversity of behaviour and policy interventions (see Table 2.5). In particular, it has been pointed out that a feedback loop arises when some of the information about a process is fed back to a starting point of the process, affecting that starting point. Since the response of the system affects inputs into that system, data scarcity can be a significant constraint in regional and sub-regional planning studies seeking to evaluate multi-sector dynamics (Chapman, 2014). Some studies have attempted to overcome this barrier by providing users with default data sets describing energy, water and land supplies and demands for their specific region of interest (Khan et al., 2020). Some have experimented with use of data aggregation tools and remote sensing (Manschatz et al., 2015), while others have attempted to combine traditional statistical methods and machine learning with mixed methods such as the Delphi technique (Kandill et al., 2001) to produce anonymized and interoperable data sets.

Data can play an important role in facilitating feedback when sufficiently disaggregated and when it frequently accessible. Further, it has been argued that there is a direct correlation between time horizons and data intensity; long-term forecasting requires conceptual clarity and relatively little data, while short-term

Table 2.5 The four attributes of socio-natural systems with examples of variables that models could represent as proxies for sources of uncertainty.

Attribute	Sources of uncertainty captured	Examples of variables to be represented in models
Stochastic events	Unpredictability, randomness, features arising unexpectedly.	A stochastic (as opposed to deterministic) representation of climate change impacts, technological breakthroughs, social unrest, economic crises.
Diversity of behaviour	Human behaviour (from individual behaviour to behavioural patterns and practices at level of a given population/system).	Social networks exerting group/peer pressure; attitudes towards energy and water conservation, consumer classifications, diffusion of information, social and cultural norms.
Policy interventions	Planned 'shocks' with unpredictable, particularly unintended consequences.	Standards for fuel and water efficiency, a feed-in tariff, a carbon tax, changes in levels of service provision, property valuations.
Co-evolution	Interactions and feedback loops, path dependency, emergence, temporal scales, non-linear developments.	Key relationships and interactions between the variables specified within the other attributes.

Modified from Source: Sharmina et al. (2019):21.

forecasting calls for more data-intensive methods (Bhattacharya, 2011). In this particular context, ABM approaches have been initially developed, and these approaches combine qualitative and quantitative approaches by drawing upon the use of back casting methods and practice theory (see also Poteete et al., 2010). In this connection, traditional methods such as SNA approach which commonly use households as the basic economic and social unit of analysis have been criticized for their neo-classical assumptions that; 1) households are a homogenous unit, denying the existence of variation in household systems identifiable by the region that governs the production, distribution and consumption of resources within a household; 2) according a greater weight to quantitative over qualitative analysis in support of positivist tradition, failing to capture multi-dimensional and dynamic interplay of factors such as gendered power relations and their consequences for decision- making processes relating to production, allocation and sharing of household resources which shape household rules and welfare and roles and incentives of individual household members (Kabeer, 1994; Sen, 1996; Whitehead, 1985; Young, 1993). This can prevent a more nuanced and enriched understanding of intra/inter-household relationships in relation to the natural resource management. Traditional methods tend to model consequences of policy interventions in a linear way with little regard for understanding the consequences of events propagating through the system. For example, Agrawal (2020) has explored the non-linear and rebound effects of CCTs in advancing changes in behaviour with regards to management of common property resources. This is an area where tangible improvements can be made to ABM approaches by a deliberate attemtpt to factor in process elements in the design of CCT intervention (for example, dis-aggregated household information about the impact of adopting a particular technique) or community level power dynamics (for example, what is the role of political factions in influencing adoption of Natural Resources Management (NRM) technique?) to better understand prospects for changes in behaviour that have the potential to deliver upon policy outcomes.

Besides improvements to modelling approaches, open access platforms offer opportunties to optimize upon repetive tasks in environmental research. We have previously referred to the work of Khan et al. (2020) to highlight how advances in computing processing power can support multidimensional modelling that targets policy, planning and integrative monitoring of trade-offs and feedback loops (Kurian, 2020). Understanding and modeling the interactions among energy, water and land systems at regional and subregional levels, however, presents substantial modeling and social science challenges. In this respect, the UNU-FLORES Nexus observatory is a first attempt to create an online modeling platform with the potential to effectively knit subregional energy, water and land systems together at the regional level and connect them to national and global socioeconomic and climatic forces in an internally consistent, computationally efficient, data efficient and decision-relevant way. The initiative places an emphasis on co-curation of data and models, linked databases that allow us to build upon what has already been developed by others and distance learning for co-design of research questions and pilot-testing of assessment protocols.

Khan et al. (2020) are right to point out that although multisectoral tools increasingly include representations of multiple sectors in a single analytical platform, this approach has not been widely applied at regional and subregional scales. While the Nexus Observatory has made a start (see Kurian, 2020; Twisa, 2021) there is little evidence of similar efforts to link water, energy and climate models at regional scale (https://nexusobservatory.flores.unu.edu/). The linked model method is confined to research activities and has not yet proven tractable in operational decision-making contexts that require greater simplicity and computational efficiency for stakeholder engagement and exploration of uncertainty. In this connection, Kanter et al. (2018) have argued that linked databases and the larger goal of co-curation of data and models by stakeholders would need to address the incongruities in the relationship between producers and users of information. 'There is a lack of recognition and understanding of the views of stakeholders outside the academic community as well as a lack of interest from scientists in building partnerships with stakeholders. The poor engagement often leads to tools that prescribe action instead of facilitating learning. This dynamic stems partly from the power asymmetries, with scientific knowledge often carrying outsized influence compared to local knowledge' (ibid: 13). In Chapter 3, we provide examples of how a failure to adequately engage low-level functionaries can lead to rent-seeking behaviour especially in the construction, operation and maintenance of infrastructure projects.

To overcome these challenges, we need a paradigm shift that is characterized by the following elements, several of which we will explore in greater detail in the ensuing chapters of this book:

- Clarity on what knowledge needs to be created and for which audience – advance scientific understanding or facilitate negotiation and decision-making or both? (Howarth and Monasterolo, 2016).
- Clarity on the role of boundary organizations – trust, skills and network that can guide in stakeholder selection, what aspects of behaviour, incentive structure and power dynamics needs to be changed? (Niblett, 2018).
- Clarity on boundary spanning roles – soft skills of facilitation, data synthesis, stakeholder engagement, monitoring and impact evaluation (Cvitanovic et al., 2018).
- Clarity on suite of mixed methods that can enable effective downscaling and coupling of models and modalities for identifying entry points at critical nodes of feedback loops connecting policy and scientific practice and (Kurian, 2017, 2020).
- Clarity on the principles and ethics framework that can advance a role for computing in leveraging the power of remote sensing, artificial intelligence and big data to support the development and continuous co-curation of data and models via an open-source modelling platform (Kanter et al., 2018).

6. Conclusions

As we examined earlier, blind spots in environmental governance have emerged because of discrepancies in the process of knowledge production and translation. The discrepancies revolve around the use of models to inform decisions on the environment–development nexus. Conventional approaches such as global-scale and uni-dimensional environmental models and RCTs proceed and dwell on an incomplete and flawed understanding of feedback loops in environmental policy and governance. Our review of the literature on incrementalism and common property resources points to the need to develop a mid-range theory of change based on innovations in the use of large N-case studies, open-source platform and piloting of policy instruments in environmental governance. In this book, we emphasize that there is nothing intrinsically wrong with case studies or environmental modelling. But what is missing is the articulation of necessary complimentary measures (reflected in theory, method, models and techniques for data transformation and valorization) that have been informed by a multidisciplinary scientific engagement to effectively bridge the gaps between science and policy. We refer to this body of literature encompassing a discussion of boundary conditions, boundary organizations and boundary spanning roles/skills in environmental governance as *boundary science.*

Boundary science specifically focusses on improving our understanding of conditions that advance synergies in environmental governance through cooperation between tiers of government, public sector departments and agencies and through partnerships with the private sector and community-based NRM organizations that are led by a reformed and revitalized bureaucratic apparatus. Our theory of change aims to shed light on the prospects for leaders in public agencies and community groups to actively engage in crafting rules for benefit distribution; negotiate with elites and factions; liaise with local governments, make upfront investments for operation and maintenance of critical infrastructure; resolve conflicts and manage climate-based risks (Jones, 2004; Vedeld, 2000). By playing this coordination role, leaders in community groups can provide a service to both governments and local resource users; we have previously referred to the undertaking of this role as 'co-provision' (Kurian and Dietz, 2012).

Effective co-provision is made possible by community leaders who maneuver in a milieu of social, ethno-racial and gender-based relationships and hierarchies, exercise of discretion by public officials, uncertainty in factor and product markets and risks posed by variability in climatic, soil and groundwater conditions (Long et al., 1986). We are optimistic that our analysis would extend existing theories by demonstrating (1) that a rational action framework which advocates for homogeneity of community for the sustainable NRM can lead to an overstatement of risks of elite capture of project/program benefits due to inadequate historical analysis of community power relations and (2) that heterogeneity can provide a basis for construction of social hierarchy, but its actual influence on collective action is mediated by social norms in the context of the specific trajectory of institutional development that can impinge upon both the magnitude and reach of the environment–development nexus (Johnson, 2004).

Mindful of a rationale and theoretical significance of institutional trajectories assessment in the environmental governance, we hope by now readers may be curious of how we can all 'practice' boundary science – or more precisely, the way in which we design, collect and analyze data to enhance our understanding and contribution to policy and environmental governance. Let us now turn to the next chapter where we will apply the conceptual insights derived from this chapter to elaborate upon the methodological constructs of a data light approach to monitoring the environment-development Nexus.

References

Abelson, P., 2003. Public Economics- Principles and Practice. Applied Economics-Canberra, ACT, Sydney.

ADB, WSP, 2010. Sanitation Finacne in Rural Cambodia: Review and Recommendations (Washington DC).

Agarwal, B., 2000. Conceptualising environmental collective action: Why gender matters. Camb. J. Econ. 24, 283–310.

Agrawal, A., 2020. Social assistance programs and climate resilience: reducing vulnerability through cash transfers. Curr. opin. Sustain. Sci. 44, 113–123.

Albrecht, T., Crootof, A., Scott, C.A., 2018. The water-energy-food nexus: a systematic review of methods for nexus assessments. Environ. Res. Lett. 13, 48–56. https://doi.org/10.1088/1748-9326/aaa9c6.

Allers, M., Ishemoi, L., 2011. Do formulas reduce political influence on intergovernmental grants? Evidence from Tanzania. J. Dev. Stud. 47 (12), pp1781–1797.

Baland, J., Platteau, P., 1996. Halting Degradation of Natural Resources? Is There a Role for Rural Communities. Clarendon Press, Oxford.

Banerjee, A., Duflo, E., 2011. Poor Economics- Rethinking Poverty and the Ways to End it. Penguin Books, New Delhi.

Barbier, E., 1997. Comment on "environment , poverty and economic growth" by Karl-Goran Maler. In: Annual World Bank Conference on Development Economics. The World Bank, Washington DC.

Bardhan, P., 2005. Introduction. Scarcity, conflict and cooperation - essays in the political and institutional economics of development, 1st. MIT, Massachusetts, pp. 1–28.

Bates, R.H., 1995. Social dilemmas and rational individuals: an assessment of the new Institutionalism. In: Harriss, Hunter, J.J., Lewis, C. (Eds.), The New Institutional Economics and Third World Development, pp. 27–48. London: Routledge.

Bazillian, M., Rogner, H., Howells, M., Hermann, S., Arnett Gielen, D., Steduto, P., 2011. Considering the water, energy and food nexus: towards integrated modelling approach. Energy Pol. 39 (12), 7896–7906.

Bebbington, A., 1997. Capitals and Capabilities: a framework for analyzing peasant viability, rural livelihoods and poverty. World Dev. 27 (12), 2021–2044.

Berardo, R., Lubell, M., 2016. Understanding what shapes a polycentric governance system. Publ. Adm. Rev. 76 (5), 738–751.

Bhattacharya, S., 2011. Introduction. Energy economics - concepts, issues, markets and governance, 1st. Springer, Netherlands, pp. 1–30.

Bjorkman, J., 2010. Budget support to local government: theory and practice. In: Kurian, M., McCarney, P. (Eds.), Peri-Urban Water and Sanitation Services: Policy, Planning and Method. Springer, Dordrecht, pp. 171–192.

Bleischwitz, R., Spataru, C., VanDeveer, S., Obersteiner, M., Van der Voot, E., Johnson, C., et al., 2018. Resource nexus perspectives on the united nations sustainable development goals. Nat. Sustainabil. 1, 737–743. https://doi.org/10.1038/s41893-018-0173-2.

Blokland, M., 2010. Benchmarking water services. In: Kurian, M., McCarney, P. (Eds.), Peri-urban Water and Sanitation Services- Policy, Planning and Method. Springer, Dordrecht, pp. 243–266.

Boserup, E., 1990. Economic and Demographic Relationships in Development. The John Hopkins University, Baltimore and London.

Boyne, G., 1996. Competition and local government—a public choice perspective. Urban Stud. 33 (4–5), 703–721.

Bromley, D.W., 1992. The commons, property and common property regimes. In: Bromley, D.W. (Ed.), Making the Commons Work. International Centre for Self-Governance, San Francisco.

Cai, X., Wallington, D., Shafiee-Jood, M., Marston, L., 2017. Understanding and managing the water-energy-food nexus- opportunities for water resources research. Adv. Water Resour. 111, 259–273. https://doi.org/10.1016/j.advwatres.2017.11.014.

Campbell, B., Beare, D., Bennett, E., Hall-Spencer, J., Ingram, J., Jaramillo, F., Ortiz, R., Ramankutty, N., Sayer, J., Shindell, D., 2017. Agricultural production as a major driver of the earth system exceeding planetary boundaries. Ecol. Soc. 22 (4), 8.

Chapman, S., 2014. A framework for monitoring social process and outcomes in environmental programs. Eval. Progr. Plann. 47, 45–53.

Connell, R.W., 1990. The state, gender and sexual politics: theory and appraisal. Theory Soc. 19 (5), 507–543.

Crenshaw, K.W., 2017. On Intersectionality-Essential Writings. New Press, New York.

Crutzen, P.J., Stoermer, E.F., 2000. Global change newsletter. Anthropocene 1, 17–18.

Cumming, G., Epstein, G., Anderies, J., Apetrei, C., Baggio, J., Bodin, O., Chawla, S., Clemens, H., Cox, M., Egli, L., Gurney, G., Lubell, M., Magliocca, N., Morrison, T., Muller, B., Seppelt, R., Schluter, M., Unnikrishnan, H., Villamayor-Tomas, S., Weible, C., 2020. Advancing understanding of natural resource governance using socio-ecological systems framework: a post-Ostrom Research Agenda. Curr. Opin. Sustain. Sci. 44, 26–34.

Cvitanovic, C., Lof, M., Norstrom, A., Reed, M., 2018. Building university-based boundary organizations that facilitate impacts on environmental policy and practice. PloS One 13 (9), 1–19. https://doi.org/10.1371/journal.pone.0203752.

Dasgupta, S., Deichmann, U., Meisner, C., Wheeler, D., 2005. Where is the poverty-environment nexus? Evidence from Cambodia, Lao PDR and Vietnam. World Dev. 33, 617–638. https://doi.org/10.1016/j.worlddev.2004.10.003.

Daher, B., Mohtar, R., Pistikopoulus, P., Portney, K., Kaiser, R., Saad, W., 2018. Developing socio-techno-economic-political solutions for addressing resource nexus hotspots. Sustainability 10 (512), 1–14.

Dasgupta, P., 2004. Human Well Being and the Natural Environment. Oxford University Press.

Dick, R.M., Janssen, S., Kandikuppa, R., Chaturvedi, K., Rao, Theis, S., 2018. Playing games to save water: collective action games for groundwater management in Andhra Pradesh. World Dev. 107, 40–53.

Do, P., Tian, F., Zhu, T., Zohidov, B., Ni, G., Lu, H., Liu, H., 2020. Exploring synergies in the water-food-energy nexus by using an integrated hydro-economic optimization model for the Lancang-Mekong river basin. Sci. Total Environ. 728. https://doi.org/10.1016/j.scitotenv.2020.1379960048-9697@2020.

Dombrowsky, I., Hesengerth, O., 2018. Governing the water-energy-food nexus related to hydropower on shared rivers- the role of regional organizations. Front. Environ. Sci. 6, 153. https://doi.org/10.3389/fenvs.2018.00153.

Dorward, A., Kydd, A., Morrison, J., Pouton, C., 2005. Institutions, markets and economic co-ordination: linking development policy to theory and practice. Dev. Change 36 (1), 1–25.

Duraiappah, A., 1996. Poverty and Environmental Degradation: A Literature Review and Analysis. CREED Working Paper Series (8).

Fan, S., Hazell, P., 2000. Should developing countries invest more in less favoured areas? an empirical analysis of rural India. Econ. Polit. Wkly. 1455–1465 (April).

Foucault, M., 1980. Power/Knoweldge: Selected interviews and other writings. In: Gordon, C. (Ed.), Harvester, Brighton, pp. 1972–1977.

Fukuyama, F., 2016. Political Order and Political Decay: From the Industrial Revolution on to the Globalization of Democracy. Farrar.

Gebrechorkos, S., Huelsmann, S., Bernhofer, C., 2019. Regional climate projections for impact assessment studies in East Africa. Environ. Res. Lett. 14, 044031.

Gibson, C., McKean, M., Ostrom, E., 2000. Explaining deforestation: the role of local institutions. In: Gibson, C., McKean, M., Ostrom, E. (Eds.), People and Forests- Communities, Institutions and Governance. MIT Press, Cambridge MA and London UK.

Gittinger, J., 1982. Economic Analysis of Agricultural Projects, EDI Series in Economic Development. The John Hopkins University Press, Baltimore.

Goldenberg, J., Johansson, T., Reddy, A., Williams, R., 1990. Energy for a Sustainable World. Wiley Eastern, Bangalore.

Gregory, R., 1997. Political rationality or incrementalism? In: Hill, M. (Ed.), The Policy Process-A Reader. Prentice Hill, Essex, pp. 175−191.

Harriss, F., Lyon, F., 2014. Transdisciplinary environmental research: a review of approaches to knowledge co-production. Nexus Think Piece Series, 002. ESRC Research Council, London.

Holling, C., 1978. Adaptive Environmental Assessment and Management. John Wiley and Sons, Chichester, UK.

Howarth, C., Monasterolo, I., 2016. Understanding barriers to decision making in the UK energy-food-water nexus. Environ. Sci. Pol. 61, 53−60. https://doi.org/10.1016/j.envsci.2016.03.014.

IOM, 2015. Migration, Climate Change and the Environment. International Organization for Migration.

Iyer, P., Evans, B., Cardosi, J., Hicks, N., 2005. Rural Water Supply, Sanitation and Budget Support, Guidelines for Task Teams. The World Bank, Washington DC.

Jackson, C., 1996. Rescuing gender from the poverty trap. World Dev. 24 (3), 489−504.

Jaganathan, N.V., 1989. Poverty, Public Policies and the Environment. The world Bank Environment Working Paper 24.

Jodha, N.S., 1990. Rural Common Property Resources Contributions and Crisis. Econ. Politic. Weekly 25 (26), 65−78.

Johnson, C., 2004. Uncommon ground: the poverty of history in common property discourse. Dev. Change 35 (3), 407−433.

Jones, E., 2004. Wealth based trust and the development of collective action. World Dev. 32 (4), 691−711.

Kabeer, N., 1994. Reversed Realities: Gender Hierarchies in Development Thought. Verso, London.

Kandil, M., El-Debeiky, S., Hasanien, N., 2001. Overview and comparison of long-term forecasting tehniques for a fast developing utility: Part I. Elec. Power Syst. Res. 50, 11−17.

Kanter, D., Musumba, M., Wood, S., Palm, C., Antle, J., Balvanera, P., et al., 2018. Evaluating agricultural trade-offs in the age of sustainable development. Agric. Syst. 163, 73−88 http://dx. doi.org/10.1016/j.agsy.2016.09.010.

Keirns, D., Darton, R., Irabien, A., 2016. The energy-water-food nexus. Ann. Rev. Chem. Biomol. Eng. 7, 239−262.

Khan, Z., Wild, T., Vernon, C., Miller, A., Hejazi, M., Clarke, L., MirallesWilhelm, F., Castillo, R., Moreda, F., Bereslawski, J., et al., 2020. Metis-a tool to harmonize and analyze multi-sectoral data and linkages at variable spatial scales. J. Open Res. Software 8, 10 http://dx. doi.org/10.5334/jors.292.

Kurian, M., Ardakanian, R. (Eds.), 2015. Governing the Nexus- Water, Soil and Waste Resources Considering Global Change. Springer, Dordrecht.

Kurian, M., Suardi, M., Ardakanian, R., 2016. Nexus Planning Primer- Lessons from Case Studies. United Nations University (UNU-FLORES), Dresden.

Kurian, M., Dietz, T., 2012. Leadership on the commons: wealth distribution, co-provision and service delivery. J. Dev. Stud. 49, 1532–1547. https://doi.org/10.1080/00220388.2013.822068.

Kurian, M., McCarney, P., 2010. In: Peri-urban Water and Sanitation Services - Policy, Planning and Method. Springer.

Kurian, M., Portney, G., Rappold, G., Hannibal, B., Gebrechorkos, S., 2018. Governance of the water-energy-food nexus: a social network analysis to understanding agency behaviour. In: Huelsmann, S., Ardakanian, R. (Eds.), Managing Water, Soil and Waste Resources to Achieve Sustainable Development Goals. Springer, Cham, pp. 125–147.

Kurian, M., Reddy, V., Scott, C., Alabaster, G., Nardocci, A., Portney, K., Boer, R., Hannibal, B., 2019. One swallow does not make a summer- siloes, trade-offs and synergies in the water-energy-food nexus. Front. Environ. Sci. 7 (31), 1–17. https://doi.org/10.3389/fenvs.2019.00032.

Kuhn, A., Wolpe, A., 1978. Feminism and materialism. Feminism and Materialism: Women and Modes of Production, 1st. Routledge and Kegan Paul, London, pp. 1–10.

Kurian, M., 2010. Financing the Millennium Development Goals (MDGs) for Water and Sanitation: Issues and Options. In: Kurian, M., McCarney, P. (Eds.), Peri-urban Water and Sanitation Services- Policy, Planning and Method. Springer, pp. 133–154.

Kurian, M., 2013. Leadership on the commons - wealth distribution, co-provision and service delivery. J. Dev. Stud. 49 (11), 1532–1547.

Kurian, M., 2017. The water -energy-food nexus: trade-offs, thresholds and transdiciplinary approaches to sustainable development. Environ. Sci. Policy 68, 96–107.

Kurian, M., 2018. The water-energy-food nexus and agriculture research for development: the case for integrative modeling via place-based observatories. In: Background Paper, Consultative Group on International Agriculture Research (CGIAR) Science Forum (Stellenbosch).

Kurian, M., 2020. Monitoring versus modelling water-energy-food interactions: how place-based observatories can inform research for sustainable development. Curr. Opin. Sustain. Sci. 44, 35–41.

Kurian, M., Turral, H., 2010. Information's role in adaptive groundwater management. In: Kurian, M., McCarney, P. (Eds.), Peri-urban Water and Sanitation Services- Policy, Planning and Method. Springer, pp. 171–192.

Leach, M., Mearns, R., 1996. Challenging received wisdom in Africa. In: Leach, M., Mearns, R. (Eds.), The Lie of the Land. James Curry, Oxford.

Liu, J., Hull, V., Godfray, C., Tilman, D., Gleick, P., Hoff, H., Pahl-Wostl, C., Xu, Z., Chung, M., Sun, J., Li, S., 2018. Nexus approaches to global sustainable development. Nat. Sustain. 1, 466–476. September.

Long, N., Ploeg, J.D., Curtin, C., Box, L. (Eds.), 1986. The Commoditization Debate: Labour Process, Strategy and Social Network. Wageningen Agricultural University, Wageningen.

Mannschatz, T., Buchroithner, M., Huelsmann, S., 2015. Visualization of water services in Africa: data applications for nexus governance. In: Kurian, M., Ardakanian, R. (Eds.), Governing the Nexus- Water, Soil and Waste Resources Considering Global Change. Springer, Dordrecht.

Marsh, D., 2008. The Water-Energy Nexus: A Comprehensive Analysis in the Context of New South Wales. Faculty of Engineering and Information Technology, Sydney University of Technology, Sydney. Doctoral Thesis.
Meinke, H., Nelson, R., Kokic, P., Stone, R., Selvaraju, R., Baethgen, W., 2006. Actionable climate knowledge: from analysis to synthesis. Clim. Res. 33, 101–110.
Mies, M., 1986. Patriarchy and Accumulation on A World Scale:Women in the International Division of Labour. Zed Books Ltd., London.
Mohammadpour, M., Mahjabin, T., Fernandez, J., Grady, C., 2019. From national indices to regional action: an analysis of food, water and energy security in Equador, Bolivia and Peru. Environ. Sci. Pol. 101, 291–301.
Mosse, D., 1997. The symbolic making of a common property resource: history, ecology and landscape in a tank irrigated landscape in south India. Dev. Change 28 (3), 467–504.
Mukul, et al., 2019. Rohingya Refugees and Environment. Science 364 (6436), 138.
Muldoon-Smith, K., Greenhalgh, P., 2019. Suspect Foundations: developing an understanding of Climate-related stranded assets in the global real estate sector. Energy Res. Soci. Sci. 54, 60–67.
Niblett, R., 2018. Rediscovering a sense of purpose: the challenge of western think tanks. Int. Affairs 94, 1409–1430. https://doi.org/10.1093/ia/iiy199.
Oliver, P., Marwell, G., 1993. The Critical Mass in Collective Action- a Micro-social Theory. Cambridge University Press, New York.
Ostrom, E., 1990. Governing the Commons- the Evolution of Institutions for Collective Action. Cambridge University Press.
Pearl, J., Mackanzie, D., 2018. The Book of Why: The New Science of Cause and Effect. Allen Lane, London.
Pollitt, C., Bouckeart, G., 2000. Public Sector Management: A Comparative Analysis. University Press Oxford, Oxford.
Poteete, A., Janssen, M., Ostrom, E., 2010. Working Together- Collective Action, the Commons and Multiple Methods in Practice. Princeton University Press.
Poteete, A., Ostrom, E., 2004. Heterogeneity, Group Size and Collective Action: The Role of Institutions in Forest Management. Development and Change 35 (3), 435–461.
Rockstrom, J., Noone, W., Personn, K., Chapin, A., Lambin, F., 2009. A Safe Operating Space for Humanity. Nature 461, 461–472.
Saidi, Elagib, A.N., 2017. Towards understanding the integrative approach of the water-energy-food nexus. Sci. Total Environ. 574, 1131–1139.
Salome, A., 2010. Wastewater management under the Dutch Water Boards: any lessons for developing countries? In: Kurian, M., McCartney, P. (Eds.), Peri-Urban Water and Sanitation Services- Policy, Planning and Method. Springer, Dordrecht, pp. 111–131.
Scoblic, J., Tetlock, P., 2020. A better crystal ball- the right way to think about the future. Foreign Aff. 99 (6), 10–18.
Scott, C., Albrecht, T., De Grande, R., Teran, A., Varady, R., Thapa, B., 2018. Water security and the pursuit of earth systems resilience. Water Int. 82, 1–20. https://doi.org/10.1080/02508060.2018.1534564.
Sen, G., 1994. Poverty, economic growth and gender equity - the Asian and Pacific Experience. In: Heyzer, N., Sen, G. (Eds.), Gender, Economic Growth and Poverty: Market Growth and State Planning in Asia and the Pacific. International Books, Utrecht.
Sen, G., 1996. Gender, market and states: a selective review and research agenda. World Dev. 24 (5), 821–829.

Sharmina, M., Ghanem, D., Browne, A., Hall, S., Mylan, J., Petrova, S., Wood, R., 2019. Envisioning surprises: how social sciences could help models represent 'deep uncertainty' in future energy and water demand. Energy Res. Soci. Sci. 50, 18–28.

Sivaramakrishnan, K., 1995. Situating the subaltern: history and anthropology in the subaltern studies project. J. Hist. Sociol. 8 (4), 395–429.

Squires, J., 1999. Gender in Political Theory. Polity Press, Cambridge.

Stiglitz, J., 2000. Economics of the Public Sector, third ed. W. Norton and Company, London.

Tomich, T., Lidder, P., Coley, M., Gollin, D., Dick, R., Webb, P., Carberry, P., 2018. Food and agricultural innovation pathways for prosperity. Agric. Syst. 172, 1–15. https://doi.org/10.1016/j.agsy.2018.002.

Twisa, S., 2021. Sustainability of Rural Water Supply in Sub-saharan Afria: GIT-Based Studies in East-Central Tanzania. Technical University of Dresden-UNU-FLORES, Germany, Dresden. Ph.D. Dissertation.

UN-Water, 2015. Consolidated Metadata Note from UN Agencies for SDG 6 Indicators on Water and Sanitation. UN-Water, New York, NY.

Urbinatti, A., Fontana, M., Stirling, A., Giatti, L., 2020. 'Opening up' the governance of water-energy-food nexus: towards a science-policy-society interface based on hybridity and humility. Sci. Total Environ. 744, 140950 https:doi.org/10.1016/j.scitotenv.2020.140945.

UvW, 2003. Zuiver Afvalwater 02- Bedrijfsvergelijking zuiveringsbeheer 2002. UvW.

Vedeld, T., 2000. Village politics: heterogeneity, leadership and collective action. J. Dev. Stud. 36 (5), 105–134. June.

Vivanco, D., Wang, R., Deetman, S., Hertwich, E., 2018. Unravelling the nexus- exploring the pathways to combined resource use. J. Ind. Ecol. https://doi.org/10.1111/jiec.12733 (Yale University).

Vogeler, C., Mock, M., Bandelow, N., Schroeder, B., 2019. Livestock farming at the expense of water resources? The water-energy-food nexus in regions with intensive livestock farming. Water 11, 2330, 10:3390?w11112330.

Walle, D., Gunewardena, D., 2001. Does ignoring heterogeneity in impacts distort project appraisals? An experiment for irrigation in Vientam. World Bank Econ. Rev. 15 (1), 141–164.

Waylen, K., Blackstock, K., Van Hulst, F., Damien, C., Horvath, F., Johnson, R., Kanka, R., Kulvik, M., Macleod, C., Meissner, K., Pavelescu, M., Pino, J., Primmer, E., Risnoveneau, G., Satalova, B., Ssilander, J., Spulerova, J., Suskevics, M., Uytvanck, J., 2019. Policy-driven monitoring and evaluation: does it support adaptive management of socio-ecological systems? Sci. Total Cim. Environ. 662, 372–384.

Weitz, N., Strambo, C., Kemp-Benedict, E., Nilsson, M., 2017. Closing the governance gaps in the water-energy-food nexus: insights from integrative governance. Global Environ. Change 45, 165–173. https://doi.org/10.1016/j.gloenvcha.2017.06.006.

Wester, P., Warner, J., 2002. River Basin Management Reconsidered. In: Turton, A., Henwood, R. (Eds.), Hydropolitics in the Developing World. African Water Issues Research Institute, Pretoria.

Whitehead, A., 1985. Effects of technological change on rural women: a review of analysis and concepts. In: Ahmed, I. (Ed.), Technological and Rural Women. Allen and Unwin, London, pp. 27–64.

WHO, UNICEF, 2015. Methodological Note: Proposed Indicator Framework for Monitoring SDG Targets on Drinking Water, Sanitation, Hygiene and Wastewater. World Health Organization and United Nations International Children's Education Fund, Geneva).

Wichelns, D., 2017. The water-energy-food nexus: is the increasing attention warranted, from either a research or policy perspective? Environ. Sci. Pol. 69, 113–123. https://doi.org/10.1016/j.envsci.2016.12.018.

Williams, B., 2011. Passive and active adaptive management: approaches and an example. J. Environ. Manag. 92 (5), 465–479.

World Bank, 2004. Making Service Work for the Poor. World Development Report, Washington DC.

World Bank, 2009. Output Based Aid: A Compilation of Lessons and Best Practice Guidance (Washington DC).

WSP, PPIAF, 2009. Guide to Ring-Fencing of Local Government -Run Water Utilities. Water and Sanitation Program, Washington DC.

WSP, 2008. Developing Effective Billing and Collection Practices Field Note No. 2. Water and Sanitation Program, New Delhi.

Yang, E., Wi, S., Ray, P., Brown, C., Khalil, A., 2016. The future nexus of the Brahmaputra river basin: climate, water, energy and food trajectories. Global Environ. Change 37, 16–30.

Young, K., 1993. Planning Development with Women: Making a World of Difference. Macmillan Press Ltd, London.

Zwareveen, M., Meinzen-Dick, R., 2001. Gender and property rights in the commons: examples of water rights in South Asia. Agri. Human Val. 18 (1), 11–25.

CHAPTER

A data light approach to monitoring the environment-development Nexus

3

1. Introduction

Our urgent call for more active engagement with the governance aspect of environmental science emerged from the realization that global public goods research was finding it difficult to inform agricultural development. A critical assessment of research and development interventions will show that technical interventions especially in agriculture have underplayed the governance dimension and its role in mediating the environment-poverty Nexus. As evidence suggests, this trend is driven by a widespread, long-lasting and strongly supported assumptions within the developmental research community about the perceived benefits of technical options such as improved crop varieties, livestock breeds, agronomic practices and management models in mitigating levels of poverty (Tomich et al., 2019; Stevenson and Vlek, 2018).

We also pointed out in the previous chapter how conventional environmental modelling suffers from rigidity, which is our second point. Often, environmental modelling attempts draw upon the simplified understanding of the policy cycle which exposes their limited capacity to effectively respond to exogeneous variables that arise within the fluid and complex dynamic socio-ecological landscape (Yang et al., 2016). Thirdly, as we saw in the discussion in the previous chapter attempts to integrate the governance aspect such as game theoretic approaches have been criticized for a serious shortcoming that assumes that all participants have equal endowments and interests in pursuing a cooperative goal (Dick et al., 2018) with regards to management of environmental resources.

Against these shortcomings, the nexus approach has been developed to provide a framework for integrative modelling of trade-offs with the objective of advancing synergies in decisions on water—energy—food interactions (Kurian et al., 2019). But as we pointed out in the previous chapter, through forging robust linkages between biophysical and institutional assessments, the nexus approach will make it imperative to focus on the broader concept of environment—development nexus rather than on the limited relationship between water, energy and food. This

Boundary Science. https://doi.org/10.1016/B978-0-323-88473-0.00003-8

engagement with the broader environment-development Nexus will emphasize the importance of engaging with critical trade-offs in environmental decision making. Such an engagement will show that advancing synergies in decisions on water-energy and food would entail taking a holistic approach that involves engaging not simply with environmental resources but with their sectoral interlinkages and interdependencies with manufacturing, public health and financing and markets for environmental products and services. Furthermore, environment-development nexus will focus attention on the distributive aspect of development interventions in addition to inequal power relations associated with decision-making processes within multi-level government structures.

As Vogeler et al. (2019) point out in their analysis through providing a more multifaceted picture of governance dynamics, concepts such as thresholds and critical mass can elevate the prominence of nexus analysis in examining the role of financing, technology adoption and path dependences of institutions in sustainable development. Similarly, the nexus approach enables assessments of agricultural interventions that are more contextual through analysis of granular details. For example, by broadening the focus of longitudinal assessments to enable comparisons of organizational performance so as to identify incentive mechanisms, we can point to the non-linear, recursive and non-monotone effects of environmental policy and governance. Furthermore, improving the salience, credibility and legitimacy of public goods research, the nexus approach can play an important role in developing methodologies for co-curation of data and downscaling of environmental models with the potential to effectively monitor and evaluate the impact of interventions that pursue sustainable development (Agrawal, 2020; Gebrechorkos et al., 2019).

This chapter will elaborate on two priorities with regards to strengthening of the institutional capacity of public agencies in relation to the use of evidence in development planning. They are; 1) operational procedures and evaluation metrics in pursuit of refining the scope and modalities for data and statistics for the purpose of more effective monitoring of development outcomes and impacts and 2) integrative models that combine bio-physical and institutional perspectives (conceptual basis of environmental models). To this end, this chapter first critically examines drawbacks manifest at the levels of the project format, the project cycle and the project model in the area of conventional environmental policy monitoring and management in the context of international development that prevents coherent design and systematic and effective monitoring of environmental governance. Secondly, we will explore the results emerging from the pilot-testing of results-based financing (RBF) approaches that makes it imperative for us to focus on the broader environment–development nexus rather than on the limited bio-physical relationship between water, energy and food interactions. The third section of this chapter lays out the methodological constructs of boundary science, which are composed of two major building blocks: typology of trade-offs and index construction. To this end, we will first discuss why agency behaviour analysis is a pre-requisite for more holistic understanding of environmental governance, which is identified as the foundational methodological concept of boundary science. Secondly, we will examine five scenarios related to soil erosion, wastewater reuse and water quality presented in the form of typology of trade-offs to demonstrate variety of trade-offs between natural resources management and public service delivery in the context of environmental governance. We also investigate their methodological implications for the development of a renewed theory of change from the perspective of environmental modelling, social networks and co-provision. Thirdly, we will illustrate the rationale behind and the technical features of index construction. Through these exercises, we intend to demonstrate the role of typology construction, data aggregation, composite indices and the potential role for expert opinion in supporting data synthesis and knowledge translation for use by decision-makers. Finally, we will conclude this chapter with a discussion of 8 design principles for research with the potential to enhance synergistic monitoring of the environment–development nexus.

2. What ails approaches for monitoring of water–energy–food interactions?

In the introductory chapter of this volume, we made an important distinction between constitutional, collective choice and operational rules to emphasize the importance of institutions in sustainable development. We pointed out that by contrast, biophysical science articulations of water–energy–food interactions tend to focus on scale and underplay the importance of boundary conditions. Boundary conditions reflected in electoral and fiscal systems of each state, however, can shape rules for managing trade-offs in environmental management through its effect on thresholds to public action, a critical mass of technology and financing and networks of information exchange. Therefore, monitoring path dependencies of institutions for management of common property resources is especially important since they can help to keep the focus on *planetary* boundaries by prioritizing the critical interdependencies between environment and development. From the discussion in the previous chapter, we know that the intensity of interdependencies between environmental management and levels of poverty may vary, thereby necessitating integrative as against sectoral assessments and analysis.

Here, there are two areas where current approaches for monitoring water–energy–food interactions face challenges and thus are ineffective in our view: (1) monitoring strategies covering project and evaluation methodologies and (2) design of public budgets. In Chapter 1, we discussed the contradictions between scale and boundary conditions and its effects on monitoring of developmental interventions. Information is key to effective monitoring, and a set of assumptions usually guides the use of criteria and methods to assess whether a development program or project is achieving its goals. We emphasize the role of cognition—a set of assumptions/norms in developmental research and project operation as it is highly relevant. Dynamics within international development projects (including research initiatives) are shaped by a set of assumptions/operational norms and practices entertained by donors, scientists and public officials, which often tend to be on a collision course. As we will illustrate in the ensuing discussion, these cognitive differences are translated into ineffectual operational modality and structural conflicts evident in project design, analysis, financing cycles and assessment.

The project format—a distorted picture

The project format or design is an analytical tool that is typically used by multilateral development banks to cast proposed investment decisions and establish a framework for analyzing information from a wide range of sources (Gittinger, 1982). The project format is a good example to illustrate the implications of financial and economic appraisals undertaken in the public sector. Financial appraisals of projects estimate the profit accruing to the project operating entity or project participants. On the other hand, economic analysis measures the effect of the project on the national economy. In the case of projects aiming to improve biodiversity, for example, some form of non-market value must be estimated. This would similarly apply to water supply and sanitation projects where previous analysis has pointed to variations in cost of establishing identical treatment technologies owning to differences in input costs such as chemicals and labour depending on location (Mara, 2007). Non-market values would also have to be estimated for project impacts that will be bought and sold in markets, for example, food or energy crops (ADB, 1997).

It is crucial to note here that the conventional public policy method in valuation of environmental goods such as cost-benefit analysis has been contested based on the idea that there is a fundamental ethical flaw for public policy in adopting a narrow set of consumerist values which does not fully respond to articulate principles and ideals of citizen's values (Anderson 1993: 210). Keeping this point in mind, there are two tensions identified in relation to the project format: (1) there is a tension arising from the gap between project specific analysis and analysis of project impact for the economy, (2) there is a tension between ascribing market and non-market values for which estimates will have to be made. The project format purports to address the data challenge through making estimates extrapolated from project experience. For example, 'most countries know they must increase food production even if they cannot cite reliable figures about total production' (Gittinger, 1982:8). Over the years, multilateral aid has advanced the idea that by channeling much of the development efforts into projects, the challenge of a lack of reliable national data can be mitigated. In low-income countries, a strange paradox has commonly emerged that reflects a growing focus on project led development assistance: increasing international aid leads to a larger proportion of effective budgets being taken off-line, thereby heightening the scope of corruption and lack of accountability. While such financing appears manageable in the short term, it undercuts the long-term development of norms and institutions with the potential to address the causes of underlying food shortages in particular and poverty in general.

The project format can also explain the support for a large number of randomized control trials (RCTs) of experimental varieties of food production technologies and management models such as conservation agriculture, improved fallow (IF), alternative microdosing, integrated soil fertility management and agroforestry (Dhehibi et al., 2018). However, analyses by a premier international agriculture organization—the Consultative Group on International Agriculture Research (CGIAR)—recently found adoption rates for these technologies and management models to be consistently low and in the range of 1%—10% (Stevenson and Vlek, 2018). This rhetoric reminds us of the similar case of a vaccine that has been developed through lab experiments and trials involving a cross section of a population but fails to translate into vaccination of the larger population, thereby reducing its potential to cure a disease. Why are adoption rates for outputs of global public goods research low? The low adoption rates could arise from design flaws in RCTs. For instance, Small-scale RCTs have tried to engage with the need for proper targeting of interventions by considering the merits of running experiments with a control area and a treatment area. But it has been found that to be cost-effective, such comparisons need not be limited to making a single one-to-one comparison. In village-level randomization, eligible villages would be spread out across the landscape and enrolled into a study – often many 100 s of villages. For interventions at the level of larger administrative units (i.e., regions/countries), there are almost never enough of them to randomize across; hence, RCTs cannot be used in this way. Synthetic control methods can be applied in the contexts of these 'small N' cases, but they come with several restrictive assumptions, even if they relax the parallel trends assumption that is central to the difference in difference method of RCT design (White, 2009). Therefore, there is no shortcut to improving our understanding of boundary conditions (or environmental decision-making in less technical terms) but through a robust analysis of the institutional environment and arrangements.

The project format often proceeds on the false assumption that once a conceptual boundary has been drawn, local information on which to base the appraisal can be gathered at relatively low cost, field trials can be designed and a judgement can be made about the social and cultural institutions that might influence the outputs of projects and pace of implementation. Thus, if individual projects are not succeeding in translating the results of experiments into widespread adoption and impact at the level of the economy, can this constraint be overcome through more experiments that improve upon experimental design? Evidentially, the former option seems to be more commonly opted by the conventional public goods research so far, which have left repressive consequences (Kurian et al., 2019). Either way, we are of the view that RCTs proceed within a narrow instrumentalist mindset that assumes that change and difference among the wider population can be sufficiently well managed to advance incremental change. Through a partial experimental analysis that inherits within the modality of project formatting, we are blind-sided in terms of our understanding of the unintended consequences of policy interventions and the trade-offs involved in decision-making in the context of the broader political economy.

As you realize from our earlier discussion on RCT design, clarity on the role of boundary conditions is highly relevant and significant in the context of understanding environmental governance. The role of boundary conditions in relation to infrastructure becomes relevant when we need to understand their implications for central fiscal transfers, taxes and tariffs for specific public services. This analysis may become especially pertinent in non-bounded systems. When resources such as water, energy or food have to be procured from outside local administrative boundaries to meet demands for public services, these systems may be referred to as non-bounded systems. By contrast, boundedness of infrastructure is characterized by a situation where it is possible to procure resources such as water, energy or food to meet demands for public services locally — from within local administrative boundaries (Kurian et al., 2019). When central transfers compensate for inability of different administrative entities (such as village or municipal governments) to contribute towards construction, operation and maintenance in non-bounded systems, they may be performing an equalization function in a federal system of public finance (Stiglitz, 2000). In the absence of equalization grants, consumers of specific services may have to pay a higher tariff to support the operation of infrastructure in non-bounded systems. Depending on how equalization grants are structured their distributional impact may be different for different segments of a community. Therefore, as we pointed out earlier, however, fundamental paradigmatic change that improves our understanding of the role of feedback loops between the design and implementation of a development intervention and the uncovering of risks in policy and planning is urgently required. Paradigmatic change would entail the development of a theory of change that goes beyond an increased number of better designed RCTs. It should include more rigorous longitudinal analysis of the institutional environment and arrangements ranging from land tenure arrangements to non-market conditions in factor and product markets.

Falling between the crack of public financial cycle: The donor-funded project cycle

For a long period, international aid has operated outside of wider political and economic considerations. The cycle of donor-funded project commonly follows a five-stage sequence: identification, preparation and analysis, appraisal, implementation and evaluation (see Fig. 3.1). It is important to point out here that in several instances, the project cycle runs parallel to the budget cycle and differs also in terms of its logic. The annual government budget cycle begins with initiation of requests by agencies, often more than a year in advance of an actual budget being adopted (Bjorkman, 2010). Key actors in the preparation of budget requests (for example, ministries and departments) seek to defend their programs at the very least, but many seek to increase their requests for budgetary resources to perform their functions.

This logic differs from project cycle analysis, which focusses on comparing most cost-effective ways of achieving project outcomes by choosing from alternative

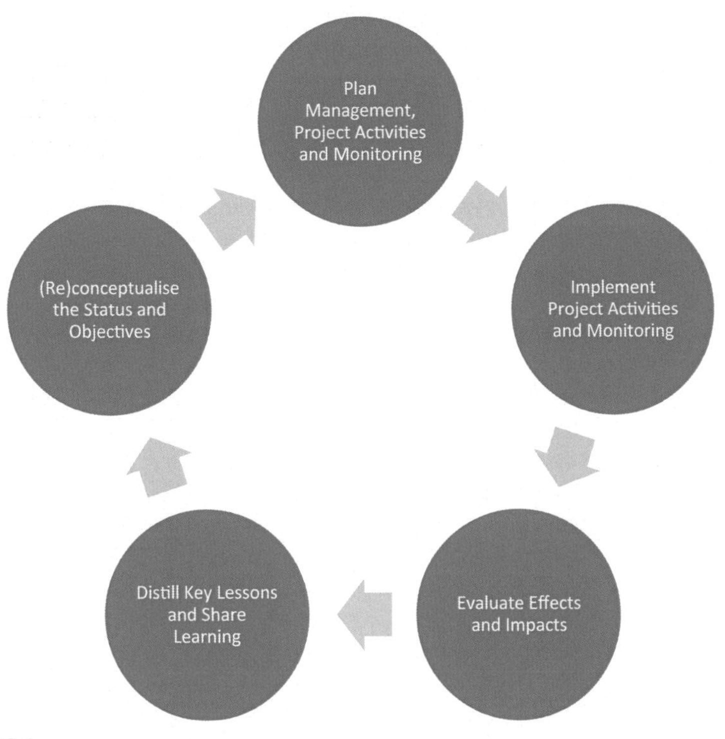

FIGURE 3.1

The adaptive management cycle.

Source: Modified from Waylen, K., et al., 2019. Policy-driven monitoring and evaluation: does it support adaptive management of socio-ecological systems? Sci. Total Environ. 662, 373–384.

options (ADB, 1997). Furthermore, project reviews of agricultural projects for example, have found that project preparation has underestimated the costs of establishing extension services that aid adoption of new practices through training programs, agronomic and livestock trials and provision of adequate credit. On the other hand, agricultural project preparation tended to be overly optimistic about the rate at which new cultivation practices would be adopted with irrigation, about the rate at which new areas would be brought under methods of improved cultivation and about the rate at which the new technology could be applied under farm conditions. In this connection, nudge studies when combined with well-designed RCTs has the potential to improve our understanding of boundary conditions in natural resources governance (Thaler, 2015). For example, cash conditional transfers (CCTs) if properly targeted can serve to nudge people to adopt water saving technologies, immunize their children or adopt high yielding seeds, thereby enhancing their potential impact on well-being.

The World Bank has in recent years attempted to overcome some of the challenges of project preparation and design by aligning project and budget cycles through the pursuit of results-based financing (RBF) strategies ranging from output-based aid (OBA) to cash conditional transfers (CCTs), Cash on Demand (COD) and budget support (Iyer et al., 2005). In the agricultural sector, results-based financing strategies draw upon lessons of earlier Bank interventions that highlighted shortcomings in project appraisal and implementation stages of the project cycle that resulted in inappropriate technology choice, inadequate attention to institutional support for extension agencies, farmer organizations, credit, information and marketing channels. Another lesson was that project appraisal and implementation failed to acknowledge the importance of administrative efficiency; projects were likely to succeed in locations where procurement delays were low and a pool of trained staff was available to undertaken project implementation (Gittinger, 1982).

Furthermore, errors that were made when translating technical assumptions into estimation of project performance could be overcome through more rigorous appraisal methods. For example, a review of appraisal methods for irrigation projects in Vietnam found that by using regional means, it was possible to reduce losses arising from erroneous project selection by between 75% and 255%. The authors point out that 'when irrigating as little as 3% of Vietnam's non-irrigated land, the savings from more rigorous data-intensive methods are sufficient to cover the full cost of extra data required for the appraisal' (Walle and Gunawardena, 2001:141).

Capturing development impacts in sustainability research: How lessons from Results-based financing (RBF) can aid a Theory of change (ToC)

Research initiatives that attempt to monitor and evaluate environment and development projects are confronted with the divergence between the different foci of the project and budget cycle. This divergence besides highlighting the tension between projects and plans, and theory and method should give us pause to think about what constitutes success and failure in research and in development practice. Let us examine these challenges further from the perspective of both of development practice and research(ToC). To begin, Fig. 3.2 offers a snapshot of intricate dynamics emerging from research-dvelopment practice nexus in this regard. As depicited in a theory of change format, production of a wide range of research outputs in the context of carbon accountability in wetlands are laid out targetting diverse stakholders at different levels and phases of the project. This figure is particularly relevant to our dicussion precisely because it does not paint the real life research challenges in the field. It rather exposes classic underlying assumptions of development research activities in the field that are carried out in a sanitized environment without much reference to local power dynamics. This lack of political economic perspective results in failing to recognize the critical role of localized institutions and norms that shape the actual research process in the field. Further, this figure is significant

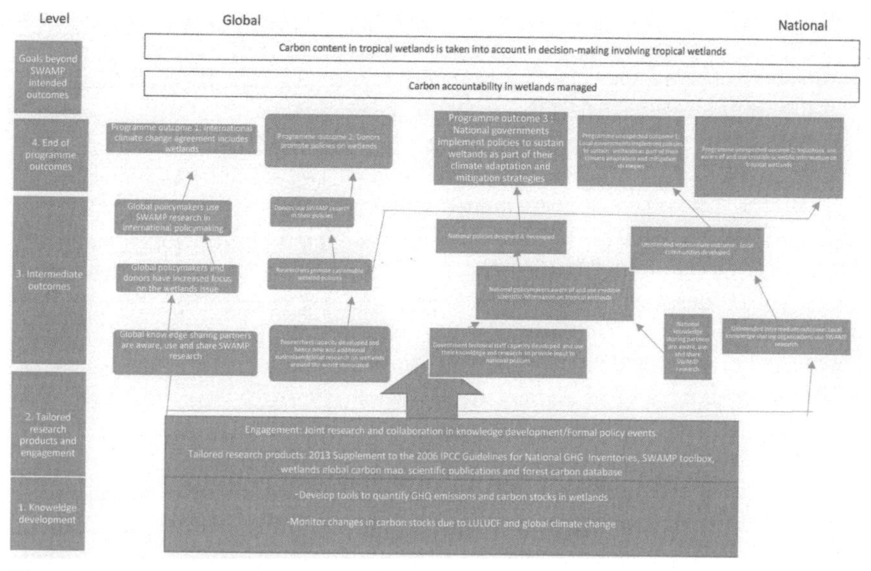

FIGURE 3.2

Theory of Change: knowledge development in the context of carbon accountability in wetlands.

Source: Adapted from Belcher, B., Claus, R., Davel, R., Ramirez, L., 2019. Linking transdisciplinary research characteristics and quality to effectiveness: a comparative analysis of five research for development projects. Environ. Sci. Pol. 101, 192–203.

as it points out to those who engaged in research and development practice the complexity and difficulty in desiging a unifying framework that can explain the conditions for success and failure of development interventions. Such a unifying framework can in turn inform the development of a theory of change that can offer insights on how to overcome the conventional assumptions of linearity and monotone effects in the design and execution of research projects.

In a similar vein, RBF initiatives such as OBA and CCTs have been developed in an attempt to engage with planning mechanisms with much more rigour and flexibility to overcome challenges related to a fixity on project design. As a result, there has been a distinct focus on project outcomes and program impact rather than on infrastructure creation alone. Different sectors—education, urban water supply and public health, lend themselves differently to institutional reform measures such as those outlined earlier.

In the case of water supply, OBA has been successfully tried in the Philippines while Brazil has succeeded with use of CCTs in improving delivery of public health services-e.g., vaccination with the objective of lowering rates of infant mortality. It is important to mention here that notwithstanding the sector, institutional reform has engaged with crucial aspects of political economy—for example, privatization of water services, tariff rates for water supply, staffing of public sector departments, engaging private sector for delivery of public services and funding for establishment, operation and maintenance of infrastructure (Batley, 2004). Differing levels of politicization of the bureaucracy in different countries have mediated the success of institutional reform measures (see Table 3.1).

Mindful that research and development practice has benefited to some degree from theoretical advancements, one of the major flaws related to architecture of both RBF and ToC relates to their limited scoping in impact assessement. Success and failure in research that is targeted on monitoring impact of developmental interventions tends to be limited to project only and ignores the broader political economy considerations within which they operate. Belcher et al. (2019), for instance, offers an interesting perspective in this regard. They use the concept of 'spheres of influence' to undertake their assessment of research undertaken by the Consultative Group on International Agriculture Research (CGIAR). They point out that project activities and outputs are mostly under the control of the project. On the other hand, project outputs, which are products, goods and services of the research product, can influence other actors and actions to contribute to positive social, economic and/or environmental changes. They distinguish between outcomes of research: changes in knowledge, attitudes and skills and relationships of actors in the system and research impact: changes in income, water discharge or water quality resulting in whole or in part from a chain of events to which the research has contributed.

The authors claim to employ a systemic approach to define a theory of change (ToC) for assessment of five CGIAR projects: management of fire and haze (Indonesia), timber extraction impact on nut production (Brazil), forest tenure reform (Peru), management of agroforestry concessions (Peru) and wetland management for climate adaptation (Indonesia). The theory of change models the change process, providing a description of how and why the project was expected to contribute to a process of change as a set of testable hypotheses. The ToC also perceives the project in terms of a sequence of steps with a probability of contributing to wider developmental outcomes.

Nevertheless, assessments of research impact on developmental outcomes suffer from three drawbacks. First, we recognize challenges incurred by budget cycles not always aligning with project cycles. In addition, non-alignment of staff transfers within public bureaucracies can also have unintended consequences for project outcomes (example, change in attitudes, knowledge and relationships) whether or not they dovetail with broader developmental outcomes. Second, hypothesis generation within a ToC project framework tends to be self-propagating. For example, if the hypothesis of research is to test the impact of a soil conservation technology, it will most likely attest to the benefits that can accrue to a farmer from using a technique in a field or controlled trial and multiply it by the number of acres of land that other farmers will adopt it on. Using current market prices, the assessment could potentially show a positive impact of adopting the improved soil conservation technology.

However, what such a theory of change overlooks is the likelihood that the farmer will not adopt the practice under real-world conditions because of the uncertainty in product prices, constraints in accessing inputs such as fertilizers or high

Table 3.1 Illustrative guide to international financing of urban sanitation services.

Sanitation service	Role specification	Institutional dimensions	Financing arrangement	International development cooperation financing instrument
Toilet construction	Municipality in consultation with ward committees	Compliance with engineering norms, targeted at poor and local tax options explored	Part municipal budget and part recovered through central transfer	Strategic lending targeted at informing budget process of municipality
Collection	Private service provider	Accountable to municipality through management contract	Tariffs/user fees	Equity fund or output based aid (OBA)
Transport	Private service provider	Accountable to municipality through management contract	Part municipal budget and part recovered through surcharge on water bill	Loan/Private sector guarantee to support implementation of build-own-transfer (BOT)/ Build-own-operate (BOO) contract
Treatment	Private service provider	Accountable to municipality through management contract; Municipality guarantees supply of water and energy	Part municipal budget and part recovered through surcharge on water bill	Guarantees for private sector investment
Reuse	Municipality and/or rural local government	Benefit-cost of reuse options ascertained	Tariffs/User fees	Grant to support design of micro-finance arrangement
Disposal	Private service provider	Compliance with environmental standards specified by relevant agency	Part municipal budget and part recovered through central transfer	Loan to support design of monitoring framework and prospective studies

Source: Kurian, M., 2010. Making sense of human-environment interactions:Policy guidance under conditions of imperfect data. In: Kurian, M., McCarney, P. (Eds.), Peri-urban Water and Sanitation Services- Policy, Planning and Method. Springer, Dordrecht, pp. 193–212.

cost of factors of production such as labour required to perform tasks such as weeding and harvesting. In Boxes 3.1-3.3, we provide some examples from OBA projects of the World Bank to emphasize the complexity involved in designing financing strategies in environmental planning. One of major drawbacks of a RBF approach is that it is blind to social traits thus subject to partisan assumptions. For example, aggregated data collected as the outcome of the OBA project interventions often presents a sanitized version of reality- that lively human experience triggered by the project interventions is tucked away, without much details offered on the beneficiaries differentiated by gender, class and ethno-racial diversity. This shortcoming is perhaps related to the fact that OBA as an evaluative methodology stems from the prime focus on pursuit of efficiency within a market-driven development paradigm that is blind to the gendered and racial hierarchal nature of political landscape associated with environmental governance. Similar to Kabeer's (1994) ciriticism against once widely praised evaluation methodology of social cost-benefit analysis, its prime focus on technological-oriented provision with a clear financial profitability target naturally makes OBA best applied in the context of interventionist and efficiency oriented project rather than a project that promotes participatory impulses with potential to enhance equity of the target group (Kabeer, 1994: 185). Therefore it is equipped to better address practical needs rather than strategic interests of the target groups/communities. Third, while not limited to OBA, attribution remains a key challenge, especially when there are multiple actors involved (with some working on similar or overlapping objectives) to ascertain whose research was instrumental in bringing about reform in critical aspects of institutions. Unless multiple research partners cooperate to achieve network effects, it is difficult to clearly ascertain how and why research impact was achieved.

Box 3.1 Country examples of OBA in practice- Uganda

In the Global Partnership on Output-Based Aid (GPOBA)-funded Water Supply in Uganda's Small Towns scheme involving small local private operators, two different output-based disbursement profiles were used: in small towns where mainly extensions from the existing system were required, a relatively "pure" OBA is used whereby private operators will be paid after connections and water service delivery. But in the green-field rural growth centres, output-based payments are phased such that 60% of the subsidies are disbursed during construction. Only 40% of the subsidies are disbursed with final connections and water delivery. It was estimated that the availability and cost of financing until delivery of output, as well as the newness of the approach, would result in either very high bids which the poor could not afford (since a portion of the costs would be borne through the tariff) or result in no bids. It is expected that in subsequent batches, the output-based disbursement profile could be made more aggressive in the rural growth centres, but this will have to await the results of output delivery in the current pilot.

Ten initial lots have all been bid out resulting in at least 20% efficiency gains overall. Three out of then towns did not require subsidies. Contract were signed in October 2008 and output delivery began in Q4 2008. As of April 2009, 310 yard taps and two PWPs are operational serving over 6000 beneficiaries. Independent verification of outputs is underway.

Source: *Adopted from The World Bank 2009:103.*

Box 3.2 Country examples of OBA in practice- Morocco

The kingdom of Morocco obtained a $7 million grant from the Global Partnership on Output- Based Aid (GPOBA) to demonstrate OBA mechanism to target sector funds and potential donor monies in support of the Government of Morocco's challenge to extend water supply & sewerage service in recently legalized informal settlements in peri-urban areas, as part of King's National Initiative for Human Development. Two international private sector incumbents (Amendis in Tangiers and Lydec in Casablanca), and a public sector incumbent (the Regie Autonome de Distribution d'Eau et d' Electricite de Meknes) are the service providers in their respective municipalities.

There are two levels of monitoring for this OBA scheme. A quarterly monitoring report is to be provided by each operator as part of their progress reports with information on: number of connections made; total number of beneficiary households; uptake ratio of beneficiary households in each eligible area; average monthly consumption per beneficiary household; average expenditure on service by beneficiary households; and the collection ratio for water bills and of connection fees.

A broader range of indicators area monitored on a yearly basis. Some area static indicators, some require regular monitoring. Examples include: average residential tariff for beneficiary households; project unit costs of house connection operator and target area as per actual incurred expenditure; and, discrepancies with estimated project costs. Additionally, service providers are encouraged to carrying out yearly service satisfaction surveys on a representative sample of about 10% of final beneficiaries.

As of April 2009, a total of 2895 water connections and 2846 sewerage connections (a little over 25% of the output targets) have been verified as delivered to pre-specification.

Source: *Adopted from The World Bank 2009:106.*

Box 3.3 Country examples of OBA in practice- Kenya

The Kenya Microfinance for Small Water Schemes Project innovatively uses output-based aid subsidies to leverage commercial finance for up to 21 small rural and peri urban piped water systems in the country. Projects were selected from a wide pool that expressed interest through a district level awareness campaign led by the Athi Water Services Board (regional asset holding government agency). To qualify for the private micro-finance bank's loans and the OBA subsidies, community projects must pass through K-Rep Bank's existing credit approval processes. The communities (organization into community associations) need to present their loan applications with considerable detail. In order to assist this process, the Global Partnership on Output-Based Aid (GPOBA) funded project has provided up-front technical assistance to communities, with considerable support from the multi-donor Water and Sanitation Program (WSP). Once the loan is approved, K-Rep Bank is responsible for loan (and thus project) monitoring and then ensures that systems are in place during operation, positively impacting the sustainability of the sub-projects.

Individual sub-projects are financed through community equity (20%) and a loan provided by K-Rep Bank (80%). If the sub-project meets the output targets – number of new connections and revenue collection-the output-based subsidy is released. The OBA subsidy covers 50% of the micro-finance loan. The remainder of the loan is repaid through tariff revenue collection. The loan provided by K-Rep is priced on a commercial basis and has a maximum tenor of 5 years. It is believed that the OBA subsidy provides a degree of comfort to the bank, and hence collateral requirements were less strenuous than usual, and the tenor was extended to 5 years over the normal 1–2 years. To provide additional security, K-Rep has arranged a USAID DCA credit guarantee for the life of the loan.

A grant provided by the Public Private Infrastructure Advisory Facility (PPAIF)is being used to develop a project development facility to assist communities in the preparation of loan applications. Further, funding from the EU Water Facility is being used to increase the scale and scop of the project to a national level. As of April 2009, construction is underway for eight of the 11 sub-projects approved by K-Rep. One sub-project was completed and verified, and two sub-projects are about to be verified. The first verified sub-projects has delivered 62 connections (24% over the output target) and the revenue from water sales was 27% more than the target.

Source: *Adopted from The World Bank 2009:104.*

3. Methodological constructs of boundary science: Studying agency behaviour in the environment–development nexus

We have hinted earlier at the role of political economy factors—why often public agencies do not respond when the data points to an acute trade-off? Why is there a deliberate institutional response when the evidence is not as compelling? (Kurian et al., 2019). For example, despite a long-standing criticism, it is still a dominant trend in the area of international development that public budgets towards infrastructure such as construction of school buildings and water treatment plants continue to rise while little attention is paid to teacher training programmes or repair and maintenance of existing treatment facilities. This conundrum highlights the great disconnect between what is appealing to science/scientific community as against what is expedient or politically possible in environmental policy and management. In response, we call for urgent recognition of the role of agency behaviour that is shaped by norms and institutions that influences the functioning of public sector bureaucracies and trigger or suppress institutional changes, an issue which has been long overlooked in environmental studies. We argue that more insights into agency behaviour are crucial to bridge the gap between science and policy and these are reasons why they are in need of greater elaboration.

First, agency behaviour analysis offers a clearer linkage between outcomes of research and impact of development interventions. Our previous experience with designing and implementing development research has convinced us that there is merit in integrating perspectives emerging from outcomes of research and impact of interventions. What this means is that if research has resulted in changes in attitudes or increased knowledge, a robust monitoring regime must attempt to attribute development impact (for example, increased income or improvements in water supply) to specific research activities or outputs. The population argument that is very often conflated to explain the causes of environmental degradation may be a good example to make our point. The population argument highlights the unintended consequences that can arise when explanations of modernization and development are taken out of their context (Giddens, 1990). Similar error can be found in the common and simplistic argument that supports experimental trials of new varieties of crops and seeds with potential to improve agricultural production.

Mosse (2005) points out that improved seeds would not solve the problems of crop production on their own. This is because seeds are not material things but an embodiment of a nexus of interacting relations that span both social and ecological domains. In other words, the prospects that improved varieties of seeds would be adopted on a large scale depend upon changes in agronomic practices, credit relations and support from professionally trained and motivated staff in extension agencies. This is the precise reason why it is critical to have States maximize utilization of well-designed and integrative policy instruments that reflect the supporting changes in agency behaviour that were responsible in bringing about holistic and sustainable development. Haskel and Westake (2020) in their new book on the intangible economy point out in this regard that States benefit from the effects of 4S': scalability, sunkeness, spillovers and synergies. Scale effects are turbo-charged with the extensive network of extension agencies (sunk costs) which can be used repeatedly to share information on weather and crop varieties and to train farmers on improved management methods. Extension agencies in turn have the potential to facilitate spillover effects of adoption of improved varieties of crops or seeds by consolidating local and regional markets and storage infrastructure. All this can in turn lead to greater synergies in support of successful pilot initiatives than can inform the design of policy instruments targeting impact on a larger scale.

Second, agency behaviour analysis enables us to draw out structural perspectives on the trajectory of changes in environmental governance. It is important to realize that the conventional research outputs such as a case study of a before and after assessment of a development intervention do little in terms of attributing the large-scale adoption of a technology or management model to changes in behaviour or knowledge. This is because such research is focussed on changes in behaviour of 'consumers or farmers' under a controlled experimental situation. Instead, what is required is to link the large-scale adoption of a technology or management model to changes in behaviour of 'agencies responsible for design, implementation, monitoring and evaluation of policy instruments': directives, notifications, guidelines, standards and circulars. The extent to which funds, functionaries and functions are reformed to address particular environmental trade-offs will reflect the extent to which agency behaviour is changing over time to address challenges such as water scarcity or poor water quality.

Francis Fukuyama (2016) in *The Origins of Political Order* points out that in a Malthusian world of growing population and limited resources, states were critical to ensuring extensive economic growth by increasing the legitimacy of systems of property rights and taxation that would ensure a steady source of revenue. Fukuyama's also restates Esther Boserup's hypothesis regarding the role of technological innovation under population growth. Fukuyama stresses the relationship between population growth and innovation is not a linear one; rather when population growth relative to resources is examined on a per-capita basis, the prospects for innovation may be higher. Therefore, careful attention to design of tax regimes for income and wealth distribution can serve to support population movements with positive effects on levels of innovation (Picketty, 2020; Oates, 1972).

Our third point is related to the benefits of agency behaviour analysis. As we discussed in Chapter 2, the study of agency behaviour encompasses the boundary conditions covering features of (1) budgetary and electoral systems, (2) horizontal and vertical dispersion of authority within the framework of a multilevel governance structure, (3) the differences in types of state systems with potential to impact upon levels autonomy, (4) discretion and accountability in decisions regarding intergovernmental fiscal transfers, (5) the potential of differential rates of taxation to impact upon access to public services and (6) the role of political parties in consolidating a unified position on policy (Stiglitz, 2000; Boyne, 1996). In sum, agency behaviour analysis provides us with the analytical vehicle to systematically map out public financing structure and mechanism to identify the critical points at which shifts in behaviour and knowledge occur or are hindered related to environmental governance.

Typology of trade-offs: Natural resource management versus public service delivery

As Albrecht et al. (2018) and others argue, in the absence of fuller integrative analytical framework, merely 'adding on' perspectives from the institutional domain can leave us with an incomplete understanding of prospects for environmental management. This incomplete view could result in overemphasizing environmental risk and being overly optimistic about the role of technology and financing in advancing sustainability, the trend we are currently overwhelmed by (Kurian et al., 2019). To this end, we have come to the realization that it is particularly relevant to identify a renewed theory of change by focusing on adapting hypothesis and explanation to insights gleaned from data without being over ambitious about fitting data to dominant models of environmental change (Pearl and MacKanzie, 2018).

Renewal in scientific approach may induce serious changes in how we proceed with the business of environmental governance i.e., how we structure learning and capacity development to inform feedback on reforms of budgetary processes, design of training programs for staff of extension agencies and the pilot-testing of policy instruments such as guidelines, notifications, circulars, standards and directives. In this regard, we have discovered the importance of boundary conditions, agency behaviour and policy instruments from the previous discussion. The discussion on boundary conditions that define environmental decision-making processes highlights the effect that electoral and fiscal systems and differences in state structures can have on shaping levels of autonomy and accountability in environmental governance. This is where managing critical trade-offs between resource use and service delivery based on enhanced engagement with boundary conditions is essential to rooting out the challenge of fragmented decision-making. Then what are the exact methodological steps involved for examination of political economy considerations in environmental research? To answer this question, let us now turn to the first of two blocks of methodological construct of boundary science—typology of trade-offs that aims to enhance monitoring of the environment–development nexus.

4. Methodological construct 1: Typology of trade-offs

Typology construction can be instrumental in mapping institutional trajectories in the environment–development nexus for two reasons. First, typology creation as an approach sheds light on the aspect of agency behaviour. It allows us to systematically map out the trends and characteristics of public agency's response to a given environmental challenge through relevant parameters such as deployment of technology, finances and capacity building initiatives for their staff over a period of time. Typology of trade-offs analysis demonstrates links between the examination of technical interventions to trade-offs and typology considerations and how these links can be translated into weights for specific indicators that cover the biophysical, institutional and socioeconomic dimensions of an environmental challenge. In this way, some of the contradictions that we observed between exogeneous and endogenous factors in design and implementation encompassing project format, project cycle and project models can be addressed.

Table 3.2 offers a snapshot of typology of trade-offs and entailing conditions triggered by the common technological option. The first thing you may notice is that trade-offs in the context of agrarian change is largely between resource management and public service delivery. Furthermore, items listed under the typology considerations column suggest different socio-economic and geographical contexts where typology of trade-offs arises. In other words, a particular form of typology—food production versus food safety triggered by adaptation of fertilizer microdosing technology in urban area, for instance—could be found only one in a country or found in multiple forms within the country (regional or district variation) or across different countries, depending on the scope and scale of the research.

Typology of trade-offs is not limited to analysis in relation to technological option. It is versatile to accommodate other dimensions of the environment-development Nexus. In this connection, drifting away from the conventional modality of case study method, which focusses on the mere illustration of a phenomenon alone, typology construction also allows us to effectively identify research gaps in theoretical and empirical analysis and subsequent policy action by highlighting the context specific patterns and characteristics of institutional responses to environmental challenges. This is the second reason why typology construction is key to institutional trajectories analysis.

Table 3.2 Typology of trade-offs in the context of global public goods research.

CGIAR technology option	Example of trade-off	Typology considerations
Fertilizer microdosing	Food production versus food safety	Rural–urban/agroecology/water endowed/bounded energy systems/ climate stressed
Integrated soil fertility management	Soil erosion versus urban water supplies	Rural–urban/agroecology/water endowed/bounded energy systems/ climate stressed
Conservation agriculture	Agricultural productivity versus diversification of income	Rural–urban/agroecology/water endowed/bounded energy systems/ climate stressed
Agroforestry	Food production versus sustainable sources of energy	Rural–urban/agroecology/water endowed/bounded energy systems/ climate stressed
Alternate wet –drying	Environmental sustainability versus stabilization of demand for farm labour	Rural–urban/agroecology/water endowed/bounded energy systems/ climate stressed

Source: Kurian, M., Reddy, V., Scott, C., Alabaster, G., Nardocci, A., Portney, K., Boer, R., Hannibal, B., 2019. One swallow does not make a summer- siloes, trade-offs and synergies in the water-energy-food nexus. Front. Environ. Sci. 7 (32), 1–17, Special Issue on "Achieving Water-Energy-Food Nexus Sustainability- A Science and Data Need or a Need for Integrated Public Policy?" (Editors: Rabi Mohtar, Jillian Cox and Richard Lawford).

Table 3.3 List of typologies of trade-offs.

Typology No.	Trade-off between	Description
Typology 1	Agricultural productivity vs soil erosion control	Soil erosion control in Laos
Typology 2	Efficiency (agricultural productivity) vs equity (income diversification)	Monitoring wastewater reuse for SDG 6.3
Typology 3	Agricultural production vs protection of water source	Monitoring access to safe water supply in rural Tanzania
Typology 4	Soil conservation vs irrigation management	Soil erosion in India
Typology 5	Water pollution vs rural livelihoods (public health and livestock risks)	Wastewater management under urbanization in India

Keeping Table 3.2 as the background, we have selected five typologies (see Table 3.3) on the three common environmental issues of wastewater, soil erosion and water quality below, to demonstrate concrete example of analysis on the institutional responses to environmental challenges. They illustrate typologies of institutional trade-offs to environmental challenges in relation to gaps in theoretical framework that were identified by application of different research methodology to study these environmental challenges. Findings of this exercise are vital as they indicate what lessons we can learn from the unsuccessful policy and research interventions and areas where future policy research is most relevant to identify generalizable principles that can guide design of development programs in the future. Let us examine each typology now.

A. Typology 1—soil erosion control in Laos

Assessing trade-offs between productivity and conservation under the privatization of property legislation

In the wake of communism in the late 1970s, Laos began granting land titles to private individuals. As a consequence, private individuals began replacing shifting cultivation with non-rotation crops with an eye on maximizing yields. The Management of Soil Erosion Consortium identified these farming practices with the potential to maximize yields.[1] However, we identified there was a trade-off between agricultural productivity versus soil erosion control. We found that yields on farm plots under shifting cultivation tended to be lower. But the MSEC project by promoting yield enhancing management practices that did not distinguish between plots

[1] For the discussion of concepts and techniques for valuing nutrients in soil and water, see Drechsel et al. (2004).

with lower soil fertility and high slope could potentially offer lower aggregate benefits than what the controlled field trials were forecasting (see Table 3.3). We also found that the most promising technical options tended to favour middle- and high-income farmers, reinforcing the socio-economic disparity along with historic ethno-racial and gender hierarchical order.

The case study emphasized three benefits of prospective studies for scaling up outputs of environmental research (Kurian, 2010). First, consultations with farmer groups and field staff of parastatal agencies enhance social learning on processes of information exchange, capacity building and project cycle management (Biggs and Smith, 2003). Second, a number of institutional issues relating to labour availability for farm operations and access to markets for agricultural inputs could be flagged even as technology trials were concluding (Scott, 1998). Third, an added benefit of a prospective study is that a methodology that incorporates perspectives on distribution of benefits and costs of technology adoption among farmers could be developed that can help monitor the potential benefits of technology adoption for historically marginalized groups in a community. In this connection, Sen (1999) makes the case for assigning evaluative weights to different components of quality of life, i.e., forest biodiversity, soil fertility or household income, and to then place the chosen weights for open public discussion. It has been pointed out that such an exercise can provide a context to the analysis by focussing on intergenerational distribution of the benefits of technology adoption for food security at household and community levels (Dasgupta, 2001).

B. Typology 2—monitoring wastewater reuse for SDG 6.3

Coupled model for benchmarking of regional achievement of global goals

This typology illustrates trade-off between efficiency (agricultural productivity) and equity (income diversification) inherent in the SDG framework. Our research covering Indonesia, Brazil and India showed that indicators currently being used by the UN on Sustainable Development Goal (SDG) 6.3—reuse/recyle of wastewater did not explicitly consider the issue of wastewater reuse (Kurian et al., 2019). Furthermore, the monitoring methodology was focussed on reporting the status of wastewater reuse and not on understanding the incentives that would make *effective* reuse possible (United Nations, 2015). We devised a monitoring methodology that would be interoperable which implies (1) the methodology could enable comparisons based on typologies of indicators in response to a policy concern that has been validated at appropriate regional scale (Kurian et al., 2019) and (2) the methodology engages scientists and non-experts to construct composite indices and facilitate data transformation and visualization in support of evidence-based decision-making.

The adoption of the Nexus framework for the research served to highlight crucial trade-offs between management of environmental resources and delivery of public services. For example, while water reuse may contribute towards water security, they may increase energy requirements and risk contamination of potable water with consequences for public health. Furthermore, the predicted reduction in demand for potable water due to implementation of reuse alternatives may be smaller than expected precisely because, for example, cost savings may drive up demand for services by consumers. The Wastewater Reuse Effectiveness Index that was developed as part of the research raised the prospect of using systematic literature reviews and expert opinion to develop and pilot-test policy instruments in pursuit of wastewater reuse.

An important question that arose as a result of the research was: what are merits of relying upon extensive surveys for a range of indicators that are not strung together in a coherent manner to be able to guide policy interventions? By hypothetically plotting countries along a continuum using the composite index instead, we were able to understand the importance of understanding local context in explaining the divergence between planetary-scale imperatives of promoting resource reuse and administrative-scale opportunities and constraints that shape both the scale and intensity of institutional responses to the SDGs. We find composite indices can be useful in structuring discussions relating to the choice of allocation and equity norms for public services that we alluded to in Chapter 2. The discussion can also make us understand how a data light monitoring paradigm can impact upon capacity building of country nodal agencies that have grown accustomed to collecting data on state of the environment and infrastructure without understanding the institutional basis for sustainable development. We found in this regard that the Delphi technique while useful in advancing new strategies to data collection and analysis also requires agency staff to be trained on how to construct typologies of trade-offs and compose expert panels.

C. Typology 3—monitoring access to safe water supply in rural Tanzania

Assessing trade-offs between agricultural intensification and soil conservation

Agriculture intensification can increase base and peak water flows and exacerbate rates of soil erosion with adverse impacts on water quality. The Water Point Mapping (WPM) project of the Ministry of Water and Irrigation in Tanzania that seeks to improve access of rural populations to water supplies through construction of hand pumps to support human, and livestock activity highlights an important trade-off between prioritizing agricultural production and protection of water source, which is the first trade-off in this context. Given that not all users of water points are agricultural producers, a second trade-off is apparent between prioritizing the interests of poor rural farmers and urban residents. A key driver for this trend is rapid land cover and land use change which has resulted from conversion of woodland into settlements (Twisa, 2021) which highlights a classic political economy challenge. By framing the water quality challenge devoid of a political economy framework (with scarcity of data to assess the water-related ecosystem services), however, the study concludes that mapping of ecosystem services is crucial to monitoring land-use change assessment of the interaction between land, water and hydrological cycle.

The analysis highlights how in the absence of a unifying framework that considers the institutional environment research tends to focus on partial solutions. This is evident from one of the recommendations of the study, which is to promote agroforestry systems through tree planting in the hope that it would improve water quality and ensure productivity of biophysical resources. The recommendation highlights a major shortcoming of environmental research, which is to focus on technical options (such as agroforestry) without understanding the institutional environment (policy and land tenure structures) and institutional arrangements (markets and capacity of extension agencies responsible for information on crop prices, seeds and fertilizers) that will eventually determine adoption rates for specific technical options that aim to mitigate critical enviromental trade-offs. It is also important to point out here that given the great diversity of biophysical conditions (soil types, drainage and slope characteristics, cropping patterns and seasonal variations in rainfall and temperature), socioeconomic (labour and food habits) and institutional capacity across Tanzania, one model of water quality management will not work. This is why a water quality monitoring regime while serving to provide a unifying framework should at the same time permit comparisons that support learning and feedback into the policy process. This would enable interventions to tailor development programs to address regional variations.

It is in this connection that typologies can prove to be instrumental in mapping the biophysical landscape, identifying diversity of socioeconomic livelihood options and possible trajectories of institutional evolution that can respond to critical trade-offs in coupled human–environment systems. Typology construction will move away from focusing simply on levels of water quality risk but risk that is more broadly embedded in human–environment systems. This means that we do not stop at mapping, but we also understand the magnitude of trade-offs that would arise from focusing exclusively on either equity or efficiency considerations in management of environmental resources and services. Finally, typology construction will allow us to appreciate the institutional response that is possible given the availability of financial resources, staffing and skill set available locally. In this respect, typology construction would entail engagement with local knowledge via expert opinion surveys through recourse to mixed methods in research.

D. Typology 4—soil erosion in India

Monitoring impact of social inequality on soil conservation outcomes

This case demonstrates the trade-off between soical conservation and irrigation management. The notification on Joint Forest Management of the Government of India in the 1990s promoted deregulation of public forest lands (Dhar, 1994). The deregulation argument called for a reorganization of departments to achieve the goal of lower government deficits, promotion of efficient service delivery and lower tariffs for consumers. Public choice theory argued that greater involvement of communities in decisions regarding the management of common property resources such as forests, livestock pastures and irrigation systems would lower the costs of policing, improve environmental outcomes and enhance the performance of infrastructure. Initial studies using indicators such as land ownership or ethnicity as measures suggested that group equality would lead to greater collective action (Oakerson, 1992).

Our use of a composite index that provided a robust understanding of socio-ecological context was able to qualify the argument by suggesting that inequality may support greater collective action in the start-up phase and in the absence of alternatives (Kurian and Dietz, 2013). Through use of the composite index, we were able to demonstrate through a longitudinal study how as collective action succeeds endogenous factors (such as land sub-division among farmers) and exogenous factors (such as a regressive fiscal regime) may combine to make the distribution of endowments more homogeneous, thereby lowering the prospects for collective action over time. Interestingly, under even conditions of group inequality collective action was predicated upon a homogeneous distribution of interests among elite factions within groups that were unequal in terms of asset distribution (Vedeld, 2000; Jones, 2004).

The case study was novel because it was one of the first to draw links between changes in community level behaviour to conditions of environmental resources. For example, the extent to which individuals would stall feed their cattle would determine the extent to which saplings would regenerate, trees would grow to their full length and epiphytes would be contained in the forest ecosystem (see Table 3.4). By consciously linking stall feeding to forest condition, the study was able to critically examine the role of slope, land use change and livestock composition in explaining deforestation (Kurian and Dietz, 2013). The use of a composite index

Table 3.4 Forest regeneration in the catchment of Bharauli dam.

Parameter	Bharauli forest	Control forest belonging to community with failed collective action (thadion)
Slope (in degrees)	11	15.1
Plots with high intensity of soil erosion (%)	11.1	46.6
No. of saplings	11	8
Basal area of trees (metres)	1.81	0.35
Basal area of saplings (centimetres)	6.42 (significant at 5% confidence level)	2.78
Diversity of saplings	2.08	1.54
Density of saplings	7.72	3.53

Source: (Kurian and Dietz, 2013):1537.

enabled comparisons across groups and over time through a combination of quantitative and qualitative data collection techniques such as participatory rural appraisal, remote sensing and standard regression. The case study was also able to appreciate the role of outliers in data analysis and explore the connections between community participation and discussions on fiscal regimes—taxes, tariffs and transfers. We discovered the unintended consequences of focusing overtly on infrastructure construction at the cost of issues of equity in service delivery that ultimately defines typologies of trade-offs in environmental decision-making in a multi-scalar governance context (Mosse, 1997).

E. Typology 5—wastewater management under urbanization in India

Valuation of climate adaptation outcomes for water reuse in secondary towns

Secondary towns are experiencing the fastest rates of urbanization worldwide. Yet, compared with large cities, secondary towns tend to have less sophisticated wastewater management infrastructure in large measure because the tax base remains relatively small. In its extreme form, climate change has exacerbated the effects of lower infrastructure coverage in secondary towns in the developing world. Poor wastewater leads to contamination of sources of drinking, negative public health consequences and impacts on livestock and flooding of agricultural fields. Our research collected water samples to demonstrate that untreated wastewater was unfit for irrigation. A cost–benefit analysis indicated that if treated wastewater had the potential to increase aggregate benefits (for an area of 454 acres) by approximately USD 150,000 for the local economy annually.

This amount had the potential to reverse the negative effect (approximately USD 88,000 annually) that wastewater pollution was having on human and livestock populations in the area (Kurian et al., 2012). Considering the resource constraints of a typical town in a developing country like India, the study examined the major technical options available for treatment of wastewater in a cost-effective way to permit its use in irrigation. The study identified five options (see Table 3.5): upflow anaerobic sewage blanket (UASB), activated sewage plant (ASP), trickling filter (TF), oxidation pond (OP) and multiple oxidation ponds (MOPs). The cost–benefit analysis revealed that MOP had the highest cost–benefit and UASB had the lowest cost–benefit among all the options (Brdjanovic et al., 2004).

Despite the availability of more cost-effective options, the local government preferred the UASB option since central fiscal transfers were available for its construction. This finding reflects an important aspect of fiscal behaviour that relates to the structure of inter-governmental financing. Inter-governmental fiscal transfers are characterized by multiple sources of funding, some conditional and others unconditional. In many cases, donor funds are an important source of central fiscal

Table 3.5 Costs and benefits of wastewater treatment plant at 2.5% discount rate over 15 years (in million rupees).

Costs-benefits	UASB	ASP	TF	OP	MOP+
Capital costs (25 MLD)	74.13	80	77.5	17.5	1.00
O&M costs	1.30	2.63	1.88	1.30	0.05
Total value of benefits	10.84	10.84	10.84	10.84	10.84
Net present value (NPV)	46.42	23.60	38.42	101.66	132.64
B–C ratio	1.53	1.21	1.40	4.12	84.16

ASP, *activated sludge plant;* MOP, *multiple oxidation pond;* OP, *oxidation pond;* TF, *trickling filter;* UASB, *up-flow anaerobic sewage blanket.*
Source: Authors field survey.in (Kurian et al., 2012):58.

transfers (World Bank, 2006). As long as central transfers are not tied to accomplishment of policy outcomes such as connection of poorer households to a sewer network or piped water supply, central transfers will only encourage dependence of local governments devoid of a search for cost-effective and efficient means of service delivery.

This finding revealed a fundamental disconnect between considerations of efficiency and equity at project levels and the nature of public budgeting, which is focussed on expanding appropriations of departments for infrastructure on a yearly basis. The more items that remain untied and off-line from the main budget, the more difficult it is to hold agencies accountable for expenditure that is not sufficiently mitigating the trade-offs between addressing an environmental challenge (water pollution) and rural livelihoods (lowering public health/livestock risks for poorer consumers). The study was able to demonstrate the use of a range of valuation methods such as contingent valuation and replacement cost through use of quantitative and qualitative methods (Reddy and Kurian, 2010).

5. Trade-off analysis: Implications for the study of institutional trajectories in the environment-development Nexus

Before we begin our discussion on the analysis, there are two key concepts which deserve clarification from the theoretical point of view. First it is related to 'endogeneity versus exogeneity'. Scholars of common property resources have posited that endogenous issues of group heterogeneity and size and scale can be influential in determining collective action outcomes (see Young, 1995; Martin, 1995). Others have contested whether factors such as financing and technology are exogenous to cooperation (Snidal, 1995). In this connection, we argue that boundary science must find ways of narrowing the gap between exogeneous influences and endogeneity in the assessment of institutional trajectories. Boundary science can provide a unifying framework that will effectively align research questions and design with program and policy concerns.

In this regard, efforts were previously made by the International Forestry Resources and Institutions (IFRI) research project initiated by Elinor Ostrom at Indiana University to bridge the gap by building an international 'database' to support decision-making on the causes and likely strategies for reversing deforestation (Gibson et al., 2000). The three goals of the project were to (1) enhance interdisciplinary knowledge of institutional trajectories with specific reference to forest stewardship, (2) provide a methodology to ground-truth aerial data and spatially link forest use to deforestation and reforestation and (3) improve assessment capabilities of participating countries in the global south.

The project emphasizes a number of insights as to how the divergence between exogeneity and endogeneity in research can be addressed: (1) if one depends on a model that is too narrow in scope when designing and implementing policy, then the outcomes may be counter intentional or counter-intuitive, (2) IFRI searched for ways to link research talent in each country and provide opportunities to collect information, build models and to influence policy change which is as necessary as tree planting to combat deforestation and (3) IFRI pursued a strategy of starting with small-scale studies that identify the simplest concepts in which a process occurs so that particular processes can be studied on-site, starting with the most observable (read exhibiting extreme trade-offs) and moving to more global strategies (read based on cross-country comparisons) (IFRI, 1997).

The second concept which requires attention is 'cost-effective prototypes of coupled models'. As the IFRI experience suggests, assessments of institutional trajectories can be both time-consuming and rely heavily on continuous funding. Future strategies must pursue cost-effective approaches since funding and travel opportunities are likely to become vastly constrained in a post-pandemic world. Fortunately, there are a number of experiences that we can draw upon to pursue a data light approach to studying institutional trajectories. For instance, Dietz et al. (2001) were able to use a caloric terms of trade approach to model normal livestock heard size to ensure food security under drought-like conditions. The approach considers market conditions, agricultural productivity and relative sufficiency offered by exclusive reliance on a cereal versus a reliance on a strategy that combines a cereal-livestock-based survival strategy. International literature reviews when combined with regional/local expert opinion can help to identify threshold parameters for construction of caloric terms of trade that can support comparative drought assessments.

Similarly, Pincus (1996) developed a 'possessions score' to overcome difficulties of ascertaining the veracity of data on income and wealth distribution in household agrarian studies. The approach seeks to address the challenge of under-reporting by using proxy indicators of income and wealth: consumer durables, conditions of the house or dwelling and small and large means of production. These approaches offer several insights on applications of a data light approach to monitoring institutional trajectories based on context specification and analysis of comparable typologies of rural livelihoods. When combined with research on typologies and composite indices, nudge studies and RCTs such innovations in methodology can offer an exciting opportunity for refining hypotheses based on engaging with pilot-testing of policy instruments (Thaler, 2015; Banerjee and Duflo, 2011). More recently, others such as the World Resources Institute (WRI) have begun addressing the issue of cost-effectiveness by proposing the conduct of online RCTs.

Bear in mind these two points discussed above. Now, let us turn to a discussion on typology analysis which are two-fold. We have to make it clear here that all these typologies are generated from original research, which were carried out without institutional analysis considerations. Against this, the purpose of the first analysis is to illustrate missing features of institutional trajectory analysis hidden within the selected study by identifying blind spots (theoretical gaps) and creating typology based on the relevant institutional trajectory issues (typology considerations) which were ignored or inadequately addressed by the given choice of research methods. For easy reference, Table 3.6 summarizes typology based on the type of research methods utilized and areas where the theoretical gaps identified within each of five typologies are described before. The table also includes information on weights allocation method. In typology 3, we can surmise that if institutional factors were considered besides agroforestry, other options could also be recommended by the analysis. For example, how can CCT schemes be targeted to induce a change in rates of agricultural intensification in rural Tanzania? In typology 5, we can similarly surmise that if OBA instruments could be designed, local governments could be incentivized to explore cost-effective climate adaptation options that promote water reuse. These two examples show us how blind spots emerge in governance and how scientific analysis can effectively address them in the pursuit of sustainable development. As you see in the last column, we have also included the analysis of the possible methods to identify respective indicators in each context, highlighting in the process theoretically relevant issues of endogeneity/exogeneity and the need for coupled models with the potential to reduce costs of monitoring.

At a glance, all of typologies present endogeneity/exogeneity as the gaps in theoretical framework except for 2—SDG: Global Model of Water Reuse. This is because global models of water reuse are focussed on understanding the trends in different regional contexts without necessarily understanding the incentives required to improve the situation by way of institutional reform. Let us examine some of the methodological constructs emerging from each of the individual typologies. Laos case (typology 1) illustrates trade-off between agricultural productivity and conserving soil resources. Logically speaking, the choice of certain theoretical framework shapes the scope and features of one's understanding of the environmental challenges and hence determines the subsequent choice of research methods. As shown, a gap in the theoretical framework, which does not consider trends in regional crop prices and farm labour costs (we express this in the table by the term endo/exogeneity), is translated into a choice of research method applied in this study—RCT. The RCT is simply focussed on the yields of individual technical options such as IF but does not consider the wider political economy factors such as

Table 3.6 Typology of research methods and gaps in theoretical framework.

Typology no.	Research method	Trade-offs	Gaps in theoretical framework (blind spots)	Typology considerations (issues which typology is constructed around)	Allocating weights for monitoring Indicators(weights allocation methods): Case studies versus expert opinion
1	Randomized control trial	Agricultural productivity vs soil conservation	Endogeneity—exogeneity	Regional changes in product prices	Expert opinion surveys are cost-effective
2	Global model of water reuse	Agricultural productivity vs diversification of income	Cost-effective prototypes of coupled models	Regional tariffs and taxes for water and infrastructure	Expert opinion surveys are cost-effective
3	Downscaled model of water quality	Soil erosion control versus rural water supplies	Endogeneity—exogeneity	Regional changes in product prices	Expert opinion surveys combined with case studies
4	Participatory rural appraisal	Food production versus sustainable sources of energy	Endogeneity—exogeneity	Subdivision of land holdings and regressive fiscal regime	Longitudinal case studies
5	Coupling of water reuse model	Food production versus food safety	Endogeneity—exogeneity	Financing norms for public infrastructure undercut viability of low-cost technical options for wastewater treatment	Expert opinion surveys combined with case studies

changes in regional crop pricing and rigidity of land tenure reform process in post-communist era that would determine their adoption in non-experimental conditions. Furthermore, the nature of typology considerations defines how we would choose indicators for monitoring the impact of a technical intervention on policy outcomes. We find in this case that expert opinion that draws upon existing literature and expertise can prove to be a more cost-effective way of ascertaining how to allocate weights for indicators such as regional prices for crops.

In this connection, the Delphi technique together with the systematic use of expert opinion offers opportunities for a data light approach for assessment of institutional trajectories (Hsu and Sandford, 2007). Project appraisal methods such as internal rate of return, benefit–cost analysis and sensitivity analysis do not offer solutions a priori. Here, future multidimensional and policy-oriented models may be built around two contested critical issues we have discussed earlier. The endogeneity versus exogeneity and cost-effectiveness of coupled models are both important from the perspective of monitoring the environment–development nexus. By pursuing such a strategy, we envision two sets of benefits; firstly, we may be able to generate interesting hypothesis that may be rewarding theoretically but could also inform regional policy design, implementation and program evaluation. Secondly, we may also be able to fine-tune global models by identifying regional-specific indicators that lend themselves easily to data valorization in support of global benchmarking strategies. In our view, models that successfully conquer these issues could potentially effectively contribute to more effective assessment of institutional trajectories along the continuum of the environment–development nexus.

In the case of the example of a global methodology for water reuse (Typology 2), we find that a critical issue is the trade-off between income diversification and agricultural productivity. A gap in the theoretical framework (cost-effective prototypes of coupled models) led to a choice of research method that does not consider the role of regional taxes and tariffs on water and infrastructure. Accordingly, expert opinion that draws upon existing literature and expertise can prove to be a more cost-effective way of ascertaining how to allocate weights for indicators such as regional taxes and tariffs. In our case of watershed management in India (Typology 4), we find that longitudinal case studies may be important to identify cost-effective models for monitoring the changes in structural factors such as land sub-division and their impact on policy and environmental outcomes. In other situations, as in the case of water reuse in India, it may be prudent to combine expert opinion surveys with case studies to build coupled models that can effectively monitor the impact of infrastructure construction on policy outcomes as they relate to public health. These examples show how a research approach that thinks in terms of typologies of trade-offs has much to offer in terms of building a unifying theoretical framework by addressing the twin issues of endogeneity–exogeneity and construction of coupled models in environmental research. Furthermore, a framework that thinks in terms of typologies of trade-offs can also contribute towards lowering the costs of monitoring developmental interventions.

What are other features of institutional trajectory analysis can we draw from these five typologies? In the previous chapter, we have discussed how syneriges emerge when three elements of critical mass, thershold and siloes combine to produce certain instiutional and environmental outcomes. Then, what are distinctive features of the instituional and environmental synergies? Is there one hegemonic standard that we all have to aspire for? or is it rather an issue of principle sharing that as long as we embrace its principle, diversity in the conditions of institutional and environmental synergies is more pragmatic? In this regard, the conditions of institutional and environmental synergies are best illuminated by embedding the discussion related to non-monotone, non-linear and recurrsive aspects of synergies, which is our second focus of typology analysis. Here it is crucial to distinguish between simplistic and realistic senarios. In a simplistic scenario or in a view of the convetional policy assumptions, institutional and environmental outcomes would demonstrate features characterized by the singular/ standardized voice or response of consumers/users (monotone), the interaction between environmental resources and institutions are sequential and in fixed course (linear), ensured by well-coordinated and regularly updated institutional feedback mechanism (recursive). In reality, however, institutional and environmental outcomes often exhibit diverse voices (non-monotone), erratic course (non-linear) and ill-coordinated/weak updating feedback mechanism (non-recursive) features. Subsequently, one could argue that what is ideal or close to the best state of affairs in this context could be combination of non-monotone, non-linear and recursive features.

In this connection, it is important to stress however that state of non-monotonic, non-linear and non-recursive which reflects a common situation in reality, is not necessarily considered repressive or a drawback. This is because, apart from the benefits of enhancing the functional aspect of institutional resilience of the government organizations as we discussed in Chapter 1, there is a significant ethical implication that can be drawn from this condition. Given the moral obligation to ensure the equity interests of the people bestowed on policy makers, it is not only ideal but also resourceful that instituional and environmental outcomes be drawn from diverse voices and policy interventions would be implemented through instituitonal framework which is flexible and responsive to emerging changes. Thus our intention here is to demonstrate different patterns of the governments' failure to embrace the diversity, complexity and uncertainty arising from this state of affairs in the context of five different typologies by using this rhetoric.

To begin, Table 3.7 provides a snap-shot of different forms of synergies. For the sake of discussion, this table demonstrates a simplified version of state of affairs that two contrasting forms of synergies identified using conventional assumptions in policy realm and a real world scenario. As you can see, three elements of synergies represent each key area of institutional functions; critical mass for technology and financing, threshold for institutional capacity and siloes for data sharing among government organizations whereas each of conditions in the form of (non)monotone. (non)linear and (non) recursive represents diversity, time/phase-bound sequencing and organizational characteristics of the information updating and coordination respectively. Against this background in mind, now let us examine each of five typologies selected here. Table 3.8 offers a summary of this analysis in this regard. Among all, similar features of institutional and environmental outcomes are shared among Typology 3, 4 and 5, exhibiting lack of institutional synergies featured by; 1)

Table 3.7 Synergistic pathways to public action.

Policy World View (monotone, linear, recursive)				Real-World Scenario (nonmonotone, nonlinear, nonrecursive)		
Critical Mass	**Threshold**	**Siloes**	**Conditions**	**CM**	**TH**	**SIL**
One technical option, one outcome	Skills and roles synchronized	Predetermined, unidimensional information required only	(Non)monotone diversity	Access to budget with no/ wrong technology choice; no access to budget and technology available	Skills and roles not synchronized	Multifaceted information not collected
No budget constraint for most advanced technology	Incentive and budget structure aligned	Project/budget cycle aligned	(Non)linear Time/phase bound sequencing	Budget constraints	Incentive and budget structure not aligned	Project/ budget cycle not aligned
Feedback loop between expenditure and outcomes is robust	Unintended effects of policy mitigated	Interventions are responsive to feedback	(Non)recursive Organizational characteristics related to information updating and coordination	Feedback loop between expenditure and outcomes nonexistence/ weak	Unintended effects of policy exacerbated	Interventions are not responsive to feedback

CM: critical mass, TH: thresholds and SIL: siloes

Table 3.8 Delineating institutional failure via typology analysis.

Typology No.	Critical Mass	Threshold	Siloes
Typology 1	Nonrecursive (nonexistence of feedback loop: no market for crop production due to change in trade policy)	Nonmonotone (no synchronization of skills and roles: no cross-sectorial policy coordination between areas of agricultural production and trade)	Nonmonotone (absence of multi-faceted information: no disaggregated information about conditions for sustaining crop yields; no information on labour market)
Typology 2	Nonrecursive (nonexistence of feed-back loop: government needs and the designing of SDG monitoring do not align)	Nonmonotone (no synchronization between skills and roles: limited capacity to implement SDGs monitoring framework)	Nonmonotone (absence of multifaceted information: only biophysical data collected)
Typology 3	Nonlinear (no budget constraints with technological options is available: maintenance problem)	Nonrecursive (unintended effects on policy not mitigated: lack of policy recognition on effects of agricultural intensification on the quality of water supply)	Nonmonotone (absence of multifaceted information: only engineering information about water infrastructure, no data on equity in relation to the aspect of water distribution)
Typology 4	Nonmonotone (no budget constraint but wrong technological choice: short-term technology chosen in order to continue rebuilding)	Non-linear (budget structure and incentive do not align: rent-seeking behaviour by low level functionaries)	Nonrecursive (nonexistence of feedback loop: weak accountability mechanism on government infrastructure expenditure)
Typology 5	Nonmonotone (no budget constraint but wrong technological choice: no justification required to access central fiscal transfers)	Nonlinear (budget structure and incentive do not align: low-level functionaries are not consulted in technology choice)	Nonrecursive (nonexistence of feed-back loop: no official recognition of the impacts of polluted water on health and livestock economy due to lack of data)

budget constraint combined with either with or without constraints to accessing technical options (Typology 3) or with no budget constraint but choice of wrong technical option (Typology 4 & 5); 2) institutional capacity is weak due to either incentives and budget structure do not align or as the result of government failure to mitigate un-intended effects of policy interventions (Typology 3) and 3) effective data-sharing is prevented either due to lack of Management Information Systems (MIS) that entertain multi-faceted information gathering by government agencies (Typology 4 & 5) or information is available but interventions are not responsive to feed-back from monitoring systems (Typology 3).

In sum, these three typologies demonstrated all three conditions of non-monotone, non-linear and non-recursive. Typology 1 and 2 on the other hand, both shared same features without time conditions (non-linear) and thus can be considered as outliers. This is because both these typologies were reviewed at an early stage of program implementation. However, this does not mean that institutional trajectory analysis is not feasible for studies which have not completed a full project cycle. Future assessments offer an opportunity to map institutional trajectories by using data that was collected at early stage as a benchmark. In both cases, institutional and environmental outcomes are not synchronized due to; 1) non-existent or weak feed-back loop, 2) institutional skills and roles are not synchronized with dealing with uncertainty and complexity and 3) multi-faceted information is not available to support coordinated policy making and implementation. As you can see in the table presentation, the beauty of this analysis is that it helps us to gain a deeper insight on the multi-dimensional aspect of institutional agency /behaviour- by showing how a certain policy choice and resulting behaviour identifiable in the form of institutional dis-functioning can be made apparent in the context of interconnectivity among realms of budget/technical, capacity and the information sharing status. This is one of prime reasons why trade-off analysis is resourceful in identifying more effective policy remedies to mitigate non-harmonious institutional and environmental outcomes.

6. Methodology construct 2: Composite index

One of the major benefits of typology analysis is that it provides insights for allocation of weights for monitoring indicators. The next logical step hereafter is to mathematically express the relationship between a set of variables to be utilized in the index construction. In this respect, the variables drawn from the typology can be used to model trade-offs for a given environmental challenge—for example, droughts or floods that affect or are affected by anthropogenic factors. A composite index for drought adaptation can accommodate biophysical, socio-economic and institutional factors. By assigning weights for different factors in the index through expert opinion surveys, the index can be calibrated to pilot specific institutional instruments (policy directives, guidelines, standards, notifications and circulars) (see Fig. 3.3). This would enable us to ascertain the conditions that would ignite a response in terms of designation of resources (skills, technology and finances)

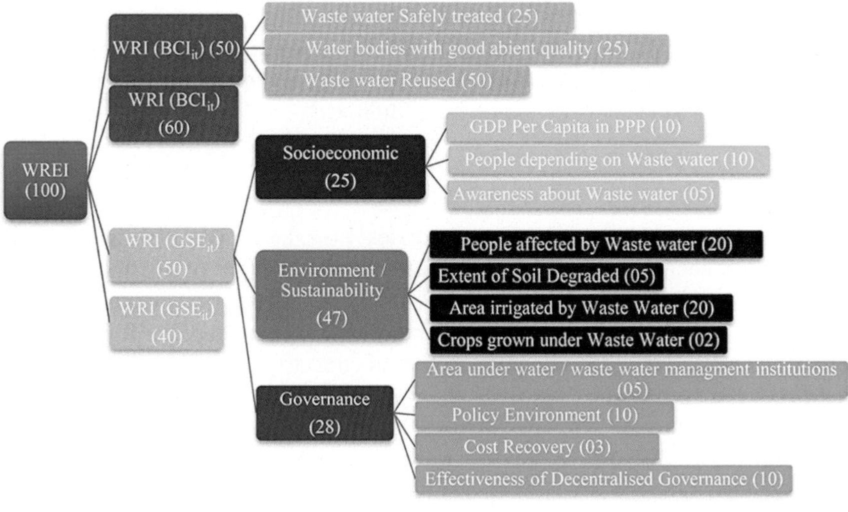

FIGURE 3.3

Typology of indicators used to model effective wastewater reuse in India. Figures in brackets are the weights based on the experience from Inda.

Kurian et al., 2019:10.

over a given period of time to enable a transition to an alternative future whereby the extreme level of trade-offs has been effectively mitigated.

Building typologies of environmental trade-offs and engaging the policy process through development and calibration of composite indices offers a way forward in the study of institutional trajectories. Let us describe how we developed an actual composite index by using one of the typologies we described earlier. Here we demonstrate the construction of two types of index. First, in Typology 4 on soil erosion in India, the composite index was constructed to predict the emergence of collective action in irrigation management. To this end, environmental trade-offs were captured by indicators of land size owned by farmers, proportion of irrigated land, livestock size and composition and household size. To be more precise, a longitudinal case study established that four indicators (1) acres of rainfed land, (2) acres of irrigated land, (3) household size and (4) size and composition of livestock were key to determining interest of individual households in managing a common property resource—an irrigation system. The distribution of household level endowments which is an integral component of the composite index was also able to illustrate the emergence of local leadership with the wealth to contribute towards routine operation and maintenance of the irrigation system. The composite index was constructed based on a secondary review of historical rainfall patterns, cereal crops calorie equivalent and milk production and calorie equivalent. The secondary review of literature was then combined with group discussions with farmers to assign weights for four variables: (1) per-acre productivity of corn/rice and wheat under nonirrigated conditions, (2) per-acre productivity of corn/rice and wheat under irrigated conditions, (3) average milk production by buffaloes in summer, monsoon and winter months and (4) average milk production by cows in summer, monsoon and winter months. Based on the assessment, the relationship among the different indicators could be mathematically expressed as follows[2]:

$$\frac{7Lr + 14Li + 2.5B + 1C + 0.5X + 0.1G}{HH\ size}$$

The formula could be used over time to assess how the structural basis of a rural community is changing in terms of household size, size and type of livestock and access to irrigation. Political economy factors such as subsidies for groundwater irrigation, expansion of nonfarm labour markets and changes in market prices for crops can all have an effect on the institutional trajectory of collective action in irrigation management. A composite index, if periodically reviewed, can be a good basis to ascertain in a cost-effective manner how developmental interventions such as an irrigation system is affecting the distribution of poverty and impacts on the environment on a scale larger than one village or irrigation system. Longitudinal analysis using a composite index can be a good guide for policy measures such as a directive on forest management or a guideline on how profits from co-management programs should be shared between community groups and governmental entities such as the forest department.

A composite index can also be useful to compare institutional responses of different regions within a country or different countries in response to a global goal. Our second example is drawn from Typology 2 which we similarly developed a composite index to monitor effective reuse of wastewater. Using the composite index, we were able to hypothetically predict the institutional trajectories of different countries and highlight some of the opportunities for capacity development. To begin, we were able to rely on secondary literature reviews of water quality and reuse to develop a prototype index to predict when effective reuse would materialize. The secondary review was able to highlight the fact that biophysical, institutional and socio-economic factors had to be considered to be able to mathematically express the potential for reuse. However, field visits to Indonesia and Brazil buttressed by feedback from countries in the Middle East made it clear that for the index to be able to support comparison it had to allow for flexibility in choice of variables under three of the broad biophysical, institutional and socio-economic headings. The choice of indicators under each of the headings and the weights that were to be

[2] The formula expresses the relationship between food security and access to irrigation in the context of a semi-arid region. Both crop productivity and milk production doubled under irrigated conditions. When applied to our study of collective action, the formula demonstrated the connection between wealth inequality (conditional upon household size) and the emergence of entrepreneurs who were more successful at managing irrigation infrastructure when compared to cooperative led management. More information about the case study can be found in Kurian M. and T. Dietz. 2012. Leadership on the commons-wealth distribution, co- provision and service delivery, Journal of Development Studies, 49, 1532–1547.

Table 3.9 WRI-BCI.

Indicator	Measure	Actual value %	Weights %	Weighted value
Waste water safely treated	%	25	25	6.25
Water bodies with good ambient quality	%	37	25	9.25
Wastewater Reuse/total wastewater#	%	20	50	10
WRI $(BCI)_{it}$		27.3*	100	25.5+

**BCI with equal weights (simple average). + BCI with differential weights. # Estimate based on the studies of various locations.*

assigned under each of the indicators had to be identified via expert opinion surveys that considered existing case studies, data and regional models.

Unfortunately, the UN SDG target 6.3 relied upon only two biophysical indicators: wastewater safely treated and water bodies with good ambient water quality. Against this, we added a third indicator—wastewater reused as a proportion of total wastewater generated to complete the biophysical component of the index. To test the application of the prototype composite index, we drew upon expert opinion surveys for India to allocate weights for different indicators of the biophysical component. We call this as WRI-BCI: Wastewater Reuse Index - Biophysical Component Index (see Table 3.9). Similarly, we relied upon expert opinion surveys to construct an institutional and socio-economic component of the index. We name this as WRI-GSE: Wastewater Reuse Index - Governance and Socioeconomic Component Index (see Table 3.10). By combining both components, the Wastewater Reuse Effectiveness Index (WREI) can be used as a tool to predict effective water reuse (see Table 3.11). The WREI offers the basis for inter-operability because it has developed a logical reasoning for why reuse is necessary and the multiple pathways (for example water reuse in agriculture or water reuse for industrial cooling) to achieving the goal. Depending on the focus the specific indicators for monitoring could differ depending on whether it is reuse in agriculture or industry that is the focus of the exercise. Further, a global monitoring strategy would accommmodate for differences in weights that are assigned to the specific issue (for example, water pricing, political awareness) even where the goal is the same to accommodate for differences in political priorities or institutional trajectories.

Composite indices can enable integrative modelling of trade-offs by incorporating information about biophysical, socio-economic and institutional dimensions of water reuse. Integrative modelling will highlight the role of political economy in decision-making by emphasizing that trade-offs are a reflection of policy and management choices that operate at global, national and local scales, which are in turn shaped by norms and agency behaviour with regards to allocation of financial and human resources and institutional capacity with implications for how effectively severity of trade-offs is mitigated. Such an approach to global monitoring of the SDGs such as target 6.3 on water reuse can present a clear picture of the constraints that various countries face and can

Table 3.10 WRI-GSE.

Component	Indicator	Measure	Actual value %	Weight %	Weighted value
Socioeconomic	Per capita GDP (PPP)	%	24	10	0.2
	People depending on waste water	%	02	10	0.2
	Awareness about waste water	%	47	05	2.35
Environment and sustainability	Population affected by water borne and water wash diseases	%	0.3	20	0.06
	Extent of soil degradation	%	29	05	1.45
	Area irrigated by waste water (potential)	%	3	20	0.6
	Crops grown under Wastewater (subsistence or high value)	% of subsistence crops	75	02	1.5
Governance	Area under water/waste water management institutions	%	22	05	1.1
	Policy environment (including water/waste water policy)	%	50	10	5
	Cost recovery	%	10	03	0.3
	Effectiveness of decentralized governance	%	31	10	3.1
WRI $(GSE)_{it}$			26.7	100	15.9

serve as an impetus for capacity building in support of normative change. Fig. 3.4 presents the hypothetical scores categorizing countries into one of the four quadrants: (H: H), (L:L), (H:L) and (L:H). This hypothetical representation offers insights on how important it is to understand local context to explain the divergence between

Table 3.11 WREI developed based on data from India.

	WRI (BCI)$_{it}$		WRI (GSE)$_{it}$		WREI$_{it}$	
Scenarios	**Normal**	**Weighted**	**Normal**	**Weighted**	**Normal**	**Weighted**
Without weights (50:50)	13.7	12.8	13.4	8.0	27.1	20.8
With weights (SI: 60:40)	16.4	15.3	10.7	6.4	27.1	21.7
With weights (SII: 70:30)	19.1	17.9	8.0	4.8	27.1	22.7

SI, Scenario one; S-II, *Scenario two.*

Source: Kurian, M., Reddy, V., Scott, C., Alabaster, G., Nardocci, A., Portney, K., Boer, R., Hannibal, B., 2019. One swallow does not make a summer- siloes, trade-offs and synergies in the water-energy-food nexus. Front. Environ. Sci. 7 (32), 1–17 Special Issue on "Achieving Water-Energy-Food Nexus Sustainability- A Science and Data Need or a Need for Integrated Public Policy?" (Editors: Rabi Mohtar, Jillian Cox and Richard Lawford).

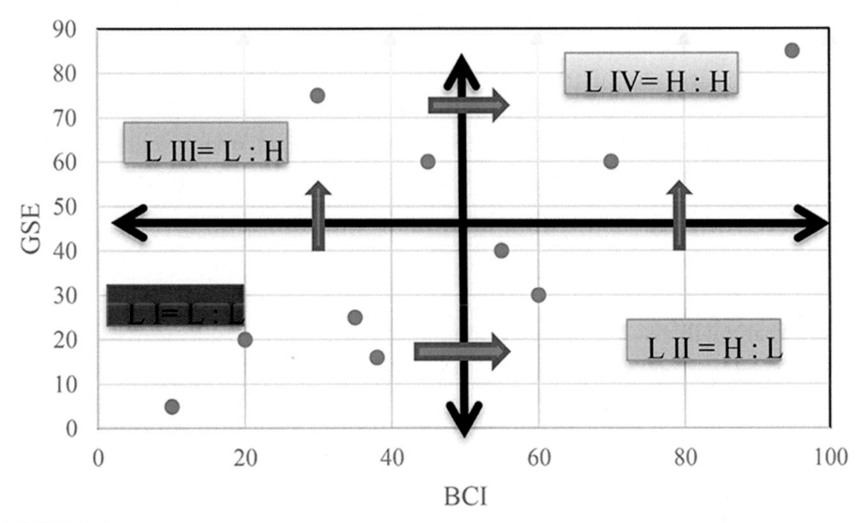

FIGURE 3.4

A hypothetical ladder for monitoring effective wastewater reuse globally.

Source: Kurian, M., Reddy, V., Scott, C., Alabaster, G., Nardocci, A., Portney, K., Boer, R., Hannibal, B., 2019. One swallow does not make a summer- siloes, trade-offs and synergies in the water-energy-food nexus. Front. Environ. Sci. 7 (32), 1–17 Special Issue on "Achieving Water-Energy-Food Nexus Sustainability- A Science and Data Need or a Need for Integrated Public Policy?" (Editors: Rabi Mohtar, Jillian Cox and Richard Lawford).

planetary-scale imperatives of promoting reuse of resources and the administrative-scale opportunities and constraints that would determine the scale and intensity of the institutional response that will ultimately drive the achievement of the SDGs.

There are several key methodological concerns that scientists pursuing assessments of institutional trajectories will have to address. These include the design and calibration of prototype indices based on typologies of environmental trade-offs. Should not case studies be driven by hypothesis that link academic discussions to a policy question? In contrast to what has been commonly believed (Kurian and Dietz, 2013), as observed in Typology 4, group heterogeneity depending on particularities of the local situation can have positive effects on cooperation specifically in the start-up phase on watershed management. This point suggests that when the particularities of a local situation are examined over time through use of a composite index, the effects of political economy on institutional trajectories become observable and as a result a range of possible policy and management interventions could be introduced. For example, taxation systems can be utilized as a tool to incentivize well-endowed individuals to undertake a co-provision role in irrigation management.

Furthermore, generating scenarios and benchmarking organizational performance can be made possible by composite indices that are interoperable. Interoperability would imply that model parameters can be altered to address regional differences without compromising the theoretical basis of the model itself. For example, in Typology 1, we were able to show that IF technical option was able to improve crop yields relative to shifting cultivation but more so for degraded plots with higher slope characteristics. Mosse (2005) is right to point out that 'removing context allows project economists to apply their own generalizing models to make predictions about the economic gains from overall yield increases'. This ability of interoperable models to minimize the gap between what they predict will happen and what actually materializes is useful in generating scenarios of institutional trajectories. This is where we believe that it is particularly important to understand a likely future scenario and think backward about the possible scenarios of how to get to that future vision. For this, it is important to have a unifying framework that envisions how IF can fit in with a broader mix of technical options to be able to achieve a future whereby trade-offs between conservation objectives and income generation have been effectively mitigated. A theory of change that articulates such a future vision will also make it clear that not all technical options with success in field trials can be accommodated by institutional trajectories that aim to address critical environmental trade-offs in decision-making. This is where composite indices have an additional attraction because they can allow us to identify conditions where institutional and technical options can work elsewhere, thereby further enhancing the allure of global public goods research.

In sum, a systematic approach to pilot-testing a mix of policy instruments to enable a transition has lessons for both science and practice. What works and how interventions fail would involve tracking composite indices- the indicators that are used and the weights that are allocated and periodically changing them to reflect changes in the political economy. From an scientific perspective this would entail targeting young scientists for training on how to construct multidimensional environmental models. This would necessitate the creation of an open-access platform, fitted with the interfaces that would allow us to co-curate data sourced from different medium (mobile, GPS, remote sensed) and from diverse sources (via linked databases) to amplify the benefits of working with typologies to advance sustainable development. This is an issue we will elaborate upon in the next chapter and in doing we would be making a marked departure from previous initiatives such as IFRI that sought to build large databases to connect the scientific community to the concerns of policy, monitoring and program evaluation. Details of open-access platform will be discussed in the following chapter.

In conclusion, based on the analysis of the five typologies and composite index, we would like to propose the following eight design principles for assessment of institutional trajectories in the environment—development nexus:

1. Understand specific local context to frame locally nuanced and sensitive research questions and identify capacity building pathways,
2. Outliers identified in analysis may offer insights for design of research focussed on understanding the conditions of institutional success and failure,
3. Composite indices can support synthesis of results from bio-physical and institutional analysis,
4. Longitudinal assessments of data via the mechanism of a composite index can enhance granularity in policy analysis,
5. Remote sensing coupled with standard assessments of forest condition or water quality can support robust downscaling of global environmental models,
6. Data visualization can enhance the power of scenario analysis and benchmarking of regional entities in discussions on incentives to improve organizational performance of sub-national governments, utilities and local authorities,
7. A creative data reuse strategy focussed on re-calibrating indicators and associated weights for each of the identified indicators can lower the costs of assessing institutional trajectories,
8. Co-curation of data and models can improve our understanding of local context and create buy-in for outputs of integrative assessments.

Descriptions of each item will become self-explanatory while we will engage in a more detailed discussion of these eight design principles in connection with open-access platform in the next chapter.

7. Conclusions

This chapter focussed on examining the methodological constructs of boundary science, which promotes enhanced monitoring of the environment—development nexus. We first began by understanding what ails monitoring approaches to water—energy—food interactions. By highlighting on the role of cognition—a set of assumptions and operational norms and practices, we discussed how the project format exacerbates some of the inconsistencies between the budget cycle and the project cycle of environment and development interventions. We concluded that one of the reasons for the poor uptake of technical interventions arises from a partial analysis that projects provides us of the broader political economy forces and the incomplete picture of feedback loops involved in public sector decision-making.

We also highlighted some of the contradictions in the logic of project budgeting and public budgeting to emphasize the need to better understand institutional context while designing, implementing, monitoring and evaluating environment–development interventions. We reviewed some of the recent attempts by the World Bank to overcome the shortcomings of a sector focus through RBF strategy that is premised on clarifying the existence of a robust feedback and linkages between environment and development interventions. We pointed out in this context that a failure to understand the policy implications brought to light by the environment-development nexus framework can result in poorly designed project appraisal and economic evaluations.

As part of a process of clarifying the methodological constructs of boundary science, we discussed the importance of studying institutional trajectories. Towards this end, we elaborated upon the foundational concept of agency behaviour analysis and why it is relevant and resourceful in studying environmental governance. We then examined two methodological constructs of boundary science typology of tradeoffs and composite indices which have the potential to foster a data light monitoring and valuation approach. We then gleaned eight design principles arising from our analysis of the five typologies covering forest management, land use management, wastewater reuse, water quality and common property resources.

The design principles highlight methodological aspects of importance to the study of institutional trajectories: longitudinal analysis, data reuse, co-curation, research framing and data generation/visualization tools and remote sensing applications. The design principles emphasize the importance of context specification in research and their applications in developing and calibrating models that can enhance the results of scenario analysis and performance benchmarking, which when viewed in connection with the mechanism of open-access platform has the potential to amplify the policy- orientation of sustainability research. [3]In conclusion, this chapter discusses ways forward and emphasizes the need to bridge gaps between exogeneity and endogeneity in environmental research, identify cost-effective strategies for building coupled models and prioritizes the role of back-casting in monitoring of environment–development interventions, an issue we take up for detailed discussion in the next chapter.

[3] As a related discussion, we should however, address here risks and limitation related to disciplinary bias arising from academic training and epistemological backgrounds of individual scientists who engage in institutional trajectories analysis in environmental governance. Since our proposed methods require scientists to work with both bio-physical and socio-economic data, it is crucial to develop carefully crafted inter-disciplinary research protocols to guide research design, data collection plan, data processing and analysis to maintain a balanced focus on natural and social sciences. Whereas the multi-disciplinary skills are indispensable, the selection of research team members should be also carried out in consultation with a perspective of whether research process could be both resorting to and enhancement of epistemological privileges of individuals. Research principles of feminist reflexivity (Mohanty, 1988; Rose, 1997) may be resourceful in this regard.

References

ADB, 1997. Guidelines for the Economic Analysis of Projects. Asian Development Bank, Manila.

Agrawal, A., 2020. Social assistance programs and climate resilience: reducing vulnerability through cash transfers. Curr. Opin. Sustain. Sci. 44, 113–123.

Albrecht, T., Crootof, A., Scott, C., 2018. The water-energy-food nexus: a systematic review of methods for nexus assessments. Environ. Res. Lett. 13, 48–56.

Anderson, E, 1993. Cost-benefit analysis, safety and environmental quality. Value in Ethics and Economics, 1. Harvard University, Massachussets, pp. 190–216.

Banerjee, A., Duflo, E., 2011. Poor Economics: Rethinking Poverty and the Ways to End it. Penguin Books, New Delhi.

Batley, R., 2004. The politics of service delivery reform. Dev. Change 35 (1), 31–35.

Belcher, B., Claus, R., Davel, R., Ramirez, L., 2019. Linking transdisciplinary research characteristics and quality to effectiveness: a comparative analysis of five research for development projects. Environ. Sci. Pol. 101, 192–203.

Biggs, S., Smith, S., 2003. A paradox of learning in project cycle managment and the role of organizational culture. World Dev. 31 (10), 1743–1757.

Bjorkman, J., 2010. Budget support to local government: theory and practice. In: Kurian, McCarney (Eds.), Peri-urban Water and Sanitation Services: Policy, Planning and Method. Springer, Dordrecht, pp. 171–192.

Boyne, G., 1996. Competition and local government—A public choice perspective. Urban Stud. 33 (4–5), 703–721.

Brdjanovic, D., Park, W., Figueres, C., 2004. Sustainable wastewater management for new communities: lessons learned from Sihwa city in South Korea. In: Proceedings of Sustainable Communities Conference, Burlington.

Dasgupta, P., 2001. Human Well-Being and the Natural Environment. Oxford University Press, UK.

Dhar, S., 1994. Rehabilitation of degraded tropical forest watersheds with people's participation. Ambio 23 (3).

Dhehibi, B., Werner, J., Qaim, M., 2018. Designing and Conducting Randamized Controlled Trials (RCTs) for Impact Evaluations of Agricultural Development Research: A Case Study from ICARDA's "Mind the Gap" Project in Tunisia. Manual and Guidelines 1. The International Centre for Agricultural Research in the Dry Areas (ICARDA), Beirut.

Dick, R.M., Janssen, S., Kandikuppa, R., Chaturvedi, K.R., Theis, S., 2018. Playing Games to Save Water: Collective Action Games for Groundwater Management in Andhra Pradesh, vol. 107. World Development, India, pp. 40–53.

Dietz, T., Nonow, A., Roba, A., Zaal, F., 2001. Pastoral commercialization: on caloric terms of trade and related issues. In: Salih, M., Dietz, T., Ahmed, A. (Eds.), African Pastoralism-Conflicts, Institutions and Government. Pluto Press, London, UK.

Drechsel, P., Giordano, M., Gyiele, L., 2004. Valuing Nutrients in Soil and Water: Concepts and Techniques with Examples from IWMI Studies in the Developing World. Research Report 82, Colombo.

Fukuyama, F., 2016. Political Order and Political Decay: From the Industrial Revolution on to the Globalization of Democracy. Farrar, New York.

Gebrechorkos, S., Huelsmann, S., Bernhofer, C., 2019. Regional climate projections for impact assessment studies in East Africa. Environ. Res. Lett. 14 (2019), 044031.

Gibson, C., McKean, M., Ostrom, E., 2000. Explaining deforestation: the role of local institutions. In: Gibson, C., McKean, M., Ostrom, E. (Eds.), People and Forests- Communities, Institutions and Governance. MIT Press, Cambridge MA and London UK.

Giddens, A., 1990. The Consequences of Modernity. Polity Press, Cambridge, UK.

Gittinger, J., 1982. Economic Analysis of Agricultural Projects, EDI Series in Economic Development. The John Hopkins University Press, Baltimore.

Haskel, J., Westake, S., 2020. Capitalism without Capital. Princeton University Press.

Hsu, C., Sandford, B., 2007. The Delphi technique: making sense of consensus. In: Practical Assessment- Research and Evaluation, vol. 12. Available online at: http://pareonline.net.

IFRI, 1997. International Forestry Resources and Institutions (IFRI) Field Manual, Version 8.5, September, Workshop in Political Theory and Policy Analysis. Indiana University, Bloomington.

Iyer, P., Evans, B., Cardosi, J., Hicks, N., 2005. Rural Water Supply, Sanitation and Budget Support, Guidelines for Task Teams. The World Bank, Washington DC.

Jones, E., 2004. Wealth based trust and the development of collective action. World Dev. 32 (4), 691–711.

Kabeer, N., 1994. Reversed Realities: Gender Hieararchies in Development Thought. Verso, London.

Kurian, M., 2010. Making sense of human-environment interactions: Policy guidance under conditions of imperfect data. In: Kurian, M., McCarney, P. (Eds.), Peri-urban Water and Sanitation Services- Policy, Planning and Method. Springer, Dordrecht, pp. 193–212.

Kurian, M., 2010. Financing the millennium development goals (MDGs) for water and sanitation: Issues and options. In: Kurian, M., McCarney, P. (Eds.), Peri-urban Water and Sanitation Services- Policy, Planning and Method. Springer, Dordrecht, pp. 133–154.

Kurian, M., Dietz, T., 2013. Leadership on the commons- wealth distribution, co-provision and service delivery. J. Dev. Stud. 49 (11), 1532–1547.

Kurian, M., Ratna Reddy, V., Dietz, T., Brdjanovic, D., 2012. Wastewater reuse for peri-urban agriculture- A viable option for adaptive water management? Sustain. Sci. 8 (1), 47–59.

Kurian, M., Reddy, V., Scott, C., Alabaster, G., Nardocci, A., Portney, K., Boer, R., Hannibal, B., 2019. One swallow does not make a summer- siloes, trade-offs and synergies in the water-energy-food nexus. Front. Environ. Sci. 7 (32), 1–17. Special Issue on "Achieving Water-Energy-Food Nexus Sustainability- A Science and Data Need or a Need for Integrated Public Policy?" (Editors: Rabi Mohtar, Jillian Cox and Richard Lawford).

Long, N., Long, A. (Eds.), 1992. Battlefields of Knowledge- the Interlocking of Theory and Practice in Social Science Research and Development. Routledge, London.

Mara, D., 2007. Selection of sanitation arrangements. In: Water Policy, vol. 9. International Water Association, The Hague.

Martin, L., 1995. Heterogeneity, linkage and commons problems. In: Keohane, R., Ostrom, E. (Eds.), Local Commons and Global Interdependence- Heterogeneity and Cooperation in Two Domains. Sage Publications, London, pp. 71–92.

Mohanty, C., 1988. Under western eyes-Feminist scholarship and colonial discourses. Fem. Rev. 30, 61–88.

Mosse, D., 1997. The symbolic making of a common property resource: history, ecology and locality in a tank-irrigated landscape in South India. Dev. Change 28 (3), 467–504.

Mosse, D., 2005. Cultivating Development: An Ethnography of Aid Policy and Practice. Vistaar Publications, New Delhi.

Oakerson, R., 1992. Analyzing the commons: a framework. In: Bromley, D. (Ed.), Making the Commons Work- Theory, Practice and Policy. Institute for Contemporary Studies Press, San Francisco.
Oates, W., 1972. Fiscal Federalism, Harcourt. Brace and Jovanovich, New York.
Pearl, J., Mackanzie, D., 2018. The Book of Why: The New Science of Cause and Effect. Allen Lane, London.
Picketty, Thomas, 2020. Introduction. Capital and Ideology, 1st. Belknap, Harvard, pp. 1–34.
Pincus, J., 1996. Class, Power and Agrarian Change. MacMillan Press, London, UK.
Reddy, V., Kurian, M., 2010. Approaches to economic and environmental valuation of domestic wastewater. In: Kurian, M., McCarney, P. (Eds.), Peri-urban Water and Sanitation Services- Policy, Planning and Method. Springer, Dordrecht.
Rose, G., 1997. Situating knowledges: positionality, reflexivities and other tactics. Prog. Hum. Geogr. 21 (3), 305–320.
Scott, J., 1998. Seeing like a State: How Certain Schemes to Improve the Human Condition Have Failed. Yale University Press, New Haven, CT.
Sen, A., 1999. Development as Freedom. Oxford University Press, USA/UK.
Snidal, D., 1995. The politics of scope: endogenous actors, heterogeneity and institutions. In: Keohane, R., Ostrom, E. (Eds.), Local Commons and Global Interdependence- Heterogeneity and Cooperation in Two Domains. Sage Publications, London, pp. 47–70.
Stevenson, Vlek, P., 2018. Assessing the Adoption and Diffusion of Natural Resource Management Practices: Synthesis of a New Set of Empirical Studies. CGIAR Standing Panel on Impact Assessment Synthesis Report, Rome.
Stiglitz, J., 2000. Economics of the Public Sector, third ed. W. Norton and Company, London.
Thaler, R., 2015. Misbehaving: The Making of Behavioural Economics. Penguin Books, New York.
Tomich, T., Lidder, P., Coley, M., Gollin, D., Dick, R., Webb, P., Carberry, P., 2019. Food and agricultural innovation pathways for prosperity. Agric. Syst. 172, 1–15. https://hdl.handle.net/10568/91278.
Twisa, S., 2021. Sustainability of Rural Water Supply in Sub-Saharan Africa: GIT-Based Studies in East-Central Tanzania., Ph.D. Dissertation. Technical University of Dresden-UNU-FLORES, Dresden, Germany.
United Nations, 2015. Consolidate Meta-Data Note from UN Agencies for SDG 6 Indicators on Water and Sanitation (New York, United Nations).
Vedeld, T., 2000. Village politics: heterogeneity, leadership and collective action. J. Dev. Stud. 36 (5), 105–134.
Vogeler, C., Mock, M., Bandelow, N., Schroeder, B., 2019. Livestock farming at the expense of water resources? The water-energy-food nexus in regions with intensive livestock farming. Water 11 (11), 2330. https://doi.org/10.3390/w11112330.
Walle, D., Gunewardena, D., 2001. Does ignoring heterogeneity in impacts distort project appraisals? An experiment for irrigation in Vietnam. World Bank Econ. Rev. 15 (I), 141–164.
Waylen, K., et al., 2019. Policy-driven monitoring and evaluation: does it support adaptive management of socio-ecological systems? Sci. Total Environ. 662, 373–384.
White, H., 2009. Theory based impact evaluations: principles and practice. In: Working Paper No. 3, International Initiative for Impact Evaluation. Global Development Network, New Delhi.
World Bank, 2006. Fiscal Decentralization in India. Oxford University Press, New Delhi.

Yang, E., Wi, S., Ray, P., Brown, C., Khalil, A., et al., 2016. The future nexus of teh Brahmaputra river basin: climate, water, energy and food trajectories. Global Environ. Chang. 37, 16–30.

Young, O., 1995. The problem of scale in human-environment relationships. In: Keohane, R., Ostrom, E. (Eds.), Local Commons and Global Interdependence- Heterogeneity and Cooperation in Two Domains. Sage Publications, London, pp. 27–46.

CHAPTER

4 Experiential learning via environmental backcasting: How open-access platforms can promote multidimensional modelling through multiple sites of engagement

1. Introduction

In the introductory chapter to this book, we argued that boundary science offers the potential to support the development of a unifying framework by elaborating upon the spanning role of boundary organizations in addressing boundary conditions of territory, authority and property. The articulation of a unifying framework can shed light on a new way of knowledge creation and translation in sustainability research. We saw that conventional environmental models and case studies have sought to build theories by resorting to assessments of technical options but underplayed the role of norms, institutions and organizations in shaping policy and management. Longitudinal case studies can potentially overcome this challenge, but the problem of research framing persists as we have demonstrated in Chapter 3. Research framing problem is a complex and multifaceted issue. Under a larger context of environment and development nexus, the dynamics of research framing is evident at the intersection of three spheres of scientific community, international development cooperation and the government policy. By engaging with the challenges of research framing it is likely that we will be able to develop a multidimensional and nuanced understanding of the significance of technology, institutional capacity/skills and financing in addressing the demands of environmental decision making.

Boundary Science. https://doi.org/10.1016/B978-0-323-88473-0.00004-X

Conventionally, the problem of research framing has overemphasised the status of biophysical resources, i.e., production of water and energy and linked cycle of resource extraction primarily through reliance upon statistical modelling. This has left us with limited scientific knowledge on socio-institutional structures and processes that play a mediatory role in producing environmental outcomes such as soil erosion, water pollution or deforestation. Consequently, the trend has brought about repressive consequences on environmental policy and governance. Inept feedback loops in environmental planning are borne by four reasons: (1) a lack of evidence-based knowledge that supports sufficient reform of agency and individual behaviour that would help in the mitigation of critical environmental trade-offs, (2) mismatch between architecture of environmental planning and mechanism of scientific research. Whereas environmental planning necessitates particular knowledge in financing and skill set such as ability to review policy trends and identify emerging needs (horizon scanning and policy priorities) and mode of policy interventions, the conduct of scientific research is not necessarily policy-driven or mindful of policy planning practices and procedures. This gap thus prevents smooth translation of research findings into effective policy interventions, (3) fundamental differences within scientific approach in terms of designing and implementing policy-relevant research in environmental governance (no unifying framework in scientific community) and (4) a lack of imagination about pathways for supporting feedback and learning within boundary organizations. These reasons that sustain a gap between research and decision making serves as the raison d taire for our urgent call for reform of scientific practice as they relate to design, implementation, monitoring and evaluation of global public goods research.

Against this background, we concluded in Chapter 2 that blind spots in environmental governance emerge because of discrepancies in knowledge translation. The discrepancies revolve around the use of environmental models, which are focussed narrowly on water—energy and food interactions that overlook the broader implications of nexus analysis in understanding the linkages, interconnections and feedback loops between resource use practices, management strategies and policy design, implementation, monitoring and evaluation. In Chapter 3, we saw how research can bridge gaps between exogeneous and endogenous factors in design and implementation by highlighting divergent perspectives between project format, project cycle and project models. Here, we emphasized the significant role of cognitive dynamics that shape our understanding of the complex socio-ecological and political economy landscape of development programming.

What then constitutes success and failure in research and development — what metrics can we employ to monitor institutional trajectories and how can composite indices assist in this process? In response to these inquires, we proposed the modus operandi (MO) of boundary science — the methodological two steps that proceed in an iterative fashion. First, drawing upon the literature reviews to creatively arrive at a characertization of a typology of trade-offs. Our analysis of five studies highlighted the importance of context specification to understand the institutional trade-offs in relation to environmental challenges and the importance of appropriate choice of research methods and framework to undertake institutional trajectory analysis. Once the typologies are constructed and monitoring indicators are identified through systematic processes, we explored the second methodological step of boundary science – composite index creation; to mathematically express the relationship between a set of variables that shape institutional responses to environmental hazards and riks.

Against this background, in this chapter, we examine the role of scientific practice in terms of curation of data and models, and a reformed set of protocols that would turbo-charge environmental models to actively translate knowledge into information that can guide decision making and enhance synergies in environmental decision making by engaging global public goods research organizations. It is in this context that the concept of an open-access platform (OAP), will be explored. We advocate for the OAP as it enables the scientific establishment to potentially benefit from lower costs of undertaking research and greater uptake of research outputs by decision-makers. Setting the eight design principles for assessment of institutional trajectories that we identified in Chapter 3 as the background, in the following Section 2, we unfold four key principles that can support the construction of OAP. The pitfalls and opportunities related to the role of OAP will be also examined. In Section 3, furthermore, we employ the example of the Belmont Forum project on cyber-enabled disaster resilience to demonstrate how the MO of boundary science can be put into practice in the design of a policy oriented research project. It emphasizes the role of global public goods research in laying the groundwork for establishment of dedicated sites for longitudinal analysis of institutional trajectories in environmental governance. Drawing upon the experience of conceptualizing the Nexus Observatory at the United Nations University, in Section 4, we dedicate our discussion to re-theorizing the environment–development nexus by highlighting the role of environmental back-casting, which may induce paradigmatic shifts in our understanding of the role of environmental planning and management in predicting and responding to the consequences of environmental hazards in the future. Section 5 summarizes the key conclusions of the chapter.

2. Open-access platform: assessment of institutional trajectories in the environment–development nexus

Engineering citizen science

So far in this book we have offered two key rationale for promoting OAP as an effective operational tool to strengthen the institutional coordination capacity in information management. They are: (a) it captures the recursive effects of institutional feedbacks in environmental governance (Chapter 1) and (b) it effectively addresses complex institutional features related to equity and inclusion in environmental decision making. As demonstrated through typology of environmental trade-offs, when it comes to the political economy of decision making, it is not so much the availability of financing, skills, and technology alone but the political will of states to deliver public services which can prove to be critical in environmental governance. Primarily because of this point, it is pertinent to move beyond the Poverty–Environment Nexus that highlights efficiency of resource use (natural/

budgetary resources) towards a framework that enables us to effectively respond also to equity related issues, not efficiency alone, under the auspices of the environmental—development nexus. To this end, we have discussed how environmental-development nexus framework is resourceful in addressing institutional trajectory analysis through the mechanism of a composite index wherein bio-physical and socio-economic features can be effectively combined (chapters 2 & 3). Similarly, concepts of citizen science and environmental back-casting, which will be discussed further in this chapter, are identified as components that can effectively enhance our understanding of political bargaining and negotiation in decisions relating to the environmental-development nexus.

We would like to emphasize that these points provide the scientific community with the rationale to welcome the OAP in principle. But as our previous experience with launching the Nexus Observatory initiative at United Nations University, suggests the scientific community is not prepared to work with the alternative but promising new ways of engaging with science. The general skepticism about the use of composite indices in applied development research is such one example. Often, scientists see an index as incomplete, less precise and a snapshot of a mathematical expression of a set of relationships (OECD, 2008). Depending on whether regression techniques are backed up by empirical data, a composite index cannot satisfy the expectations of all scientists. This is because, rather than viewing a composite index as a tool to engage in dialogue on likely scenarios of policy outcomes and enable performance benchmarking (based on a numerical expression of possible institutional trajectories), scientists tend to look at the index as an end in itself – simply as a regression exercise that can for all practical purposes be undertaken with proxies derived from secondary data. This curtails the possibility of using composite indices to identify incentives for institutional change in response to an environmental challenge, monitor the intensity of trade-offs into the future and identify critical nodes within large bureaucractic structures and processes that could provide the impetus for altering a given institutional trajectory.

Scientific bias against unconventional approaches and source of knowledge is another common challenge. While expert opinion is identified as the powerful resource, it may add to the level of scientist's despair because in their view, citizen science could compromise the very ability of science to be precise and rigorous (Harwood, 2018). How do we end up with a way to engage with science anyway? Why do we view knowledge in a hierarchical manner and where do the disparaging attitudes towards citizen science come from? Political theorist Squires (1999) explains that positivism views all knowledge claims must be empirically derived from both observation and reasoning endorsed by cognitive objectivism. The impact of positivism was indeed significant so as to trigger critical debates over positivism's defining role in furthering scientific objectivity and the theory of knowledge thereafter (Keller and Grontkowski, 1996). Grand theorists such as Harding (1987) questioned the legitimization behind granting the natural sciences with a supreme status over the social sciences precisely because of the primacy accorded to engagement with the empiricist form of cognitive objectivism by the natural scientists. Evidently, positivism has also defined how social scientists engage with knowledge production and translation, encouraging them to mime or conform to the standards of natural science methods and approaches that study the environmental domain (Harding, 1987, 89–90).

Similarly, the scientific community which engages with environmental studies tends to look for the perfect solutions often gleaned from sanitized data sets that seek to eliminate outliers and noise in the data. But very often, success and failure in development practice are explained through the life stories and wealth status of individuals within communities that are not reflected in the normalized distribution projected via data sets (Tian et al., 2018). In this connection, there is an extensive literature on epistemology that sheds light on more vibrant interpretations of connectivity between knowledge and experience from the perspective of gender and other marginalized groups in society, which is highly relevant and resourceful for engagement with sustainable development. Social scientists are as much guilty of advancing these blind spots owing to their own disciplinary orientation that places a premium on confirming hypothesis based on the role of reason. But as we saw from our soil erosion case from India in Chapter 3, hierarchy and reputation derived from distribution of wealth assets very often determines the basis of collective action and synergies in environmental management (Baland and Platteau, 1999).

In this respect, the idea of citizen science has slowly begun to be recognized within the scientific community in recent years together with a growing interest in expanding definition of scientific knowledge from the perspectives of environmental engineering (CSE, 1999), gender in development, politics and citizenship (Brodie, 1994; Cornwall, 2003; Kabeer, 2005; Razavi, 1997; Ribeiro et al., 2020) and feminist political ecology (Agarwal, 1997, 2001; Jackson, 1993; Mies and Shiva, 1993; Shiva, 1988; Sultana, 2009). In this vein, we argue that our proposition on citizen science offers a sharper conceptual lense to address operational challenges inherent in the areas of production, translation and implementation of scientific knowledge and representation[1] in the context of environment-development nexus where the role of government and citizens and its implications in redefining public policy knowledge and behaviour shaping the pubic service delivery have to be reexamined beyond the conventional structural binary and hiearchy. In this connection, we define citizen science as a body of knowledge and practice that is embedded in ways and which how scientific knowledge on the environment and technology is transformed by the cultural, institutional and normative filters of society to affect the lived experience of people and communities. Further, knowledge production as such

[1] Here it is pertinent to inquire whether the structures of representation are aligned with the needs of service delivery. There are four key questions that can guide this inquiry: (a) when claiming to be a representative, what is one representing (beliefs, constituency, interests, identities)? (b) how does one represent it (microcosm representation- age, sex, race, symbolic representation- class, principal agent based representation? (c) where does representation occur (in legislature or informally in groups)? and (d) what is the purpose of representation (to raise awareness, change laws)? (Kabeer, 2005).

entails transformative processes of scientific dialogues that engage scientific experts and practitioners to elicit feedback that is focused on improving the design, implementation and impact of development interventions. In this vein, citizen science has the potential to influence our perceptions of success and failure of developmental interventions and provide important inputs for the design of development programs (Mosse et al., 1998) such as persistent problem of refining the balancing act on efficiency-equity dichotomous challenges which are yet to be overcome. Citizen science is however predicated on an open and two-way dialogue between governments and citizens who are both participants in the design of development programs and beneficiaries (Lane, 2000). In the context of research on environment–development nexus therefore, we should be critical of the conventional normative standards related to conceptualization, methods, scope and validity of data and call for more open and creative approaches that can co-design alternative approaches of knowledge creation to acquire a more substantive basis for understanding environmental challenges. We wish to emphasize that this mindset is the first important step for engagement with boundary science, and in this spirit, we introduce the idea of OAP in the following sections of this chapter.

Open-access platform: four key components

In principle, an OAP is different from a database. From a computing perspective, OAP consists of web-based interfaces that promote the use of open-source software and tools to enable linking of databases, archiving of data, data processing tools and methods, online learning, data valorization techniques (which is different from confirming parametric consistency of models) and data transformation tools such as composite indices and scenario analysis. Unlike a database, OAP allows for co-curation of data sourced from different sources (public and private data) and using different mediums (in addition to maps, remote sensed, mobile and sensor data). When combined with the scientific process of framing the right question, OAP has much to offer in terms of transforming data for decision-making. Data transformation is enabled via software that makes it possible to aggregate data and link data from multiple data sources and points and visualize data to support scenario and benchmarking analysis, for which OAP can be guided by a strict privacy and data ethics framework. The figure summarizes how OAP draws upon citizen science perspectives to advance data valorization and web-based learning, optimize repetitive tasks in the research process and seek political buy-in for the results of scientific projects (Fig. 4.1).

Petal 1: framing success and failure in institutional analysis: the challenge of data valorization

A more durable basis for understanding the role of outliers in explanations of success and failure in institutional analysis is through construction of typologies of trade-offs. Trade-offs reflect choices given the constraints and opportunities offered

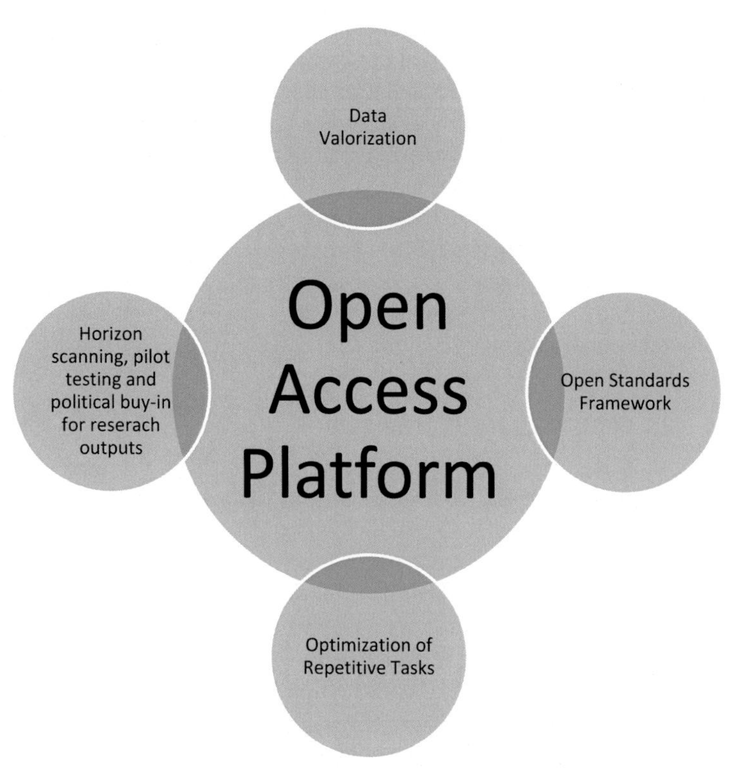

FIGURE 4.1

Key components of open-access platform.

by politics, technology and financing. By their nature, trade-offs do not point towards perfect solutions. Far from this, typologies of trade-offs do not rely on normalized data or conceptually sound scientific formulations based on analysis of existing literature. Instead typologies of trade-offs point to the best available decisions possible under the constraints imposed by political choices regarding how to advance social welfare based on considerations of imposing the lowest possible cost for the largest number of people. This form of theory building works with typologies to build a taxonomy of relationships for a set of biophysical and institutional indicators that have been knitted together through engagement with empirical conditions and relevant literature. Very often, principal investigators of research projects proceed on the belief that teams composed of social scientists would lead to better explanations of institutional success and failure. But what several others are finding instead is that common sense is not the preserve of social scientists on research teams — in fact in the absence of a well-understood framework that knits together indicators and data, scientists can continue to construct models and publish papers that confirm what we already know.[2] This is why data valorization is key to tracing if models can indeed be defended for the explanations they offer about real-world phenomenon (Sharmina et al., 2019). This is different from a process of data validation that seeks to confirm parametric consistency of model structures and veracity of data itself (Regueiro et al., 2020). Engagement with real world conditions is not a pre-requisite for data validation but it can be the sin que non of a process of data valorization.

Petal 2 — an open standards framework

The valorization of data relies upon typologies of trade-offs as a first step in the analysis. This is distinctly different from typologies of risk since the focus remains on the behaviour and interconnections between biophysical resources (for example, water and energy). But typologies of trade-offs rely on the constraints and opportunities that a given set of environmental resources offer together with the availability of the appropriate mix of technology, skill set and financing to understand the types of choices that decision-makers can possibly make in a particular political economy context. By its nature, this process of constructing typologies of trade-offs is therefore is fraught with complexity, non-linearity and deep uncertainty (Kurian, 2020). How can scientific practice embrace the complexity, non-linearity and deep uncertainty inherent in the process of tracing the policy impact of interventions in the environment—development nexus? (Regueiro et al., 2020) argue that an open standards framework in environmental research can go a long way in laying the groundwork for rigorous policy engagement. They identify five steps that encompass an open standards framework:

- Conceptualization: a vision of the desired conservation state and creation of the conceptual model of the system, interlinkages, goals, objectives, activities and expected outcomes of research and developmental threats that can constrain the achievement of research outcomes
- Plan, actions and monitoring: the desired goal, components, results chains of research project
- Implementation of activities and monitoring
- Analysis of monitoring data to assess effectiveness of proposed activities and adaptation of the conceptual model and conservation actions is possible

[2] The complexity of the environment—development nexus makes it clear that knowledge and data gaps cannot be filled by a single discipline. Regueiro et al. (2020) point out that in the absence of a unifying framework, natural scientists often focus on immediate causes of environmental decline, while social scientists study factors contributing to drivers of such decline.

- Learning and sharing the key research and policy lessons emerging from the intervention: these can include models of typology construction, models of pilot-testing composite indices and models of social learning arising from engagement with expert opinion surveys

Petal 3 — optimizing repetitive tasks in environmental research

Climate research has shown that the process of building models that engage with policy processes can be supported by mechanisms that lower the research costs involved in undertaking repetitive tasks (Kurian et al., 2018). In recent decades, a number of global and regional data sources have become available with higher spatial and temporal resolution. The process of preparing higher-resolution data involves several repetitive steps that are time-consuming. These repetitive steps can be avoided if researchers are provided access to the data without them spending time learning the processing tools related to R, Climate Data Operator (CDO) and NCO (Net CDF Operator). Similarly, an open standards framework would enable researchers to retrieve location specific minimum—maximum temperature and rainfall data to prepare a watershed-based soil budget and develop scenarios. An emphasis on overcoming repetitive tasks could also be expanded to data collection through encouragement for reuse of data (Villamayor et al., 2020). The reuse of data offers the promise of engaging with stakeholders to build regional capacity for co-curation of data to downscale global models through pilot-testing of policy instruments that better addresses local priorities. Such an approach to research would involve the development of boundary spanning skills of negotiation, online learning and horizon scanning that identifies entry points for research from the universe of possible development programs in given region (Cvitanovic et al., 2018).

Petal 4 — Pilot-testing of policy instruments to create political buy-in for research outputs

Most conventional modelling exercises start with data that is already available covering water, energy and food, generate hypothesis, test them through secondary data and conclude the obvious. Nothing new emerges both from a theoretical or methodological standpoint from research that is detached from the wider political economy (Cumming et al., 2020). Instead, a critical feature of boundary science involves the scanning of available projects, global models, RCT experiments and future policy priorities. A scanning exercise helps us identify a map of possible entry points for research that targets regional development challenges. But rather than using case studies to build snap shots of institutional trajectories that are tied to a given time period, pilot-testing policy of instruments (*directives, guidelines, notifications, standards and circulars*) offers a way of overcoming the time-constraint in policy-oriented research. This is where an OAP fitted with appropriate interfaces can enable scientists to leverage the power of linked databases, big data, remotely sensed data, virtual reality, online learning and machine learning to frame policy relevant research and incubate multi-dimensional models of environmental change and management. After all over time, public budgets and policy priorities of donors change, staff are transferred and data becomes redundant (Mohammadpur et al., 2019). This is why pilot-testing policy instruments offers an opportunity to situate policy directives within a particular political economy context of changes in norms, institutions and organizational behaviour (Gevelt, 2020). A directive or notification to pursue resource reuse, for example, reflects a broad acknowledgement on the part of decision-makers that resource reuse has potential to support income generation and promote public health. By choosing to deliberately design research that informs the implementation of an institutional instrument, one is enhancing the chances of contributing by way of inputs into processes of political negotiation and potentially creating buy-in for the outputs of research (Vasconcelos et al., 2020). In doing so, we are enhancing the potential of advancing:

- An online platform supporting research by optimizing on time and resources spent on repetitive tasks
- Social learning didactic and pedagogical approaches for studying institutional trajectories
- Co-curation of data, linked models and methods for pilot-testing of institutional instruments based on typologies of trade-offs
- Tools of data valorization and visualization that enable the use of performance benchmarks and scenario analysis by the public and private sectors and citizen groups (Ribeiro et al., 2020; Muldoon-Smith and Greenlagh, 2019)

3. Towards a model of models: longitudinal learning at multiple sites

There is no doubt that data and models are useful in generating knowledge about interactions between environmental resources such as water and energy and how they generate services such as water treatment and distribution or food processing, distribution and supply (Uden et al., 2018). But we must also understand that for such services to make any meaningful impact on livelihoods, they must operate at scale, generate spillover effects, foster synergies and/or invest sunk costs in establishing and maintaining infrastructure. Public bureaucracies offer the potential of advancing the effects of scale, spillover, synergies and sunken costs, but political expediency may limit the scope for wholehearted compromise in the context of partially understood problems. This is where *boundary science* can contribute in terms of unleashing the scale, spillover and synergistic effects of open source modelling by:

- providing a framework that would enable us to approach environmental decision-making within public bureaucracies in terms of trade-offs and less in terms of perfect solutions,
- identifying entry points for researchers to contribute in terms of integrative models that capture both biophysical and agency behaviour,

- understanding the importance of data transformation and political negotiation skills in enhancing the chances of public goods research to influence the design, implementation and monitoring policy instruments (Eftelioglu et al., 2017).

Such a project of building robust research designs that deliberately engage with public sector decision-making would enable us to appreciate the bigger picture – that environmental management is only one of several priorities confronting governments. When environmental priorities such as climate change impinge upon other developmental priorities such as income, public health or employment, there may greater scope for pilot-testing of policy instruments because of their potential to mitigate extreme trade-offs owing to the effects on other sectors. In this context, the application of the eight design principles for assessment of institutional trajectories may achieve greater traction in the context of robust feedback loops between environmental conservation and broader developmental goals.

In the ensuing discussion, we engage with the path breaking new approach of fostering regional and consortium-based research that addresses challenges of drought hazards that has implications for water and energy resources and services such as food supply. Using the example of a Belmont Forum funded research proposal, we demonstrate the applications of OAP in advancing the eight research design principles identified in Chapter 3. This Belmont proposal seeks to develop a model of models on how we can learn about institutional success and failure by unleashing the power of longitudinal analysis through engagement with multiple sites that can provide future opportunities for co-design of research, co-curation of data and online learning. While empowering local researchers, the proposal also seeks to demonstrate the network effects of aligning international partners and donors with regional developmental priorities through attention to issues of research framing and opportunities for paradigmatic change in scientific practice based on political buy-in for research outputs.

The Belmont Forum is an intentional consortium of donors and science foundations that aims to promote transdisciplinary research that provides knowledge for understanding, mitigating and responding to global environmental change. The Belmont forum exemplifies some of the core principles of Boundary Science that have been articulated in this book, namely: (a) a conviction to reform how research is framed and executed by the scientific community that would make it more accountable by responding to pressing global challenges such as climate change, (b) a conviction to address hierarchies in science so as to include the voices of young scientists drawn from multiple disciplines and development practitioners from the global south and (c) a conviction to establish multi-country research consortium around themes that have been identified as having potential to inform policy processes and structures both globally and regionally. By innovating via the Belmont principles on open data and by engaging a wide range of stakeholders through incubator and scoping workshops, the Belmont Forum in lighting a path for reform that can be beneficial for both decision makers and those with an interest in environmental science.

The Belmont Forum research consortium on cyber-enabled disaster resilience: 2020–23

Horizon scanning (research design principles 4)

Humanitarian disasters such as droughts expose approximately 10 million people in sub-Saharan Africa to food, water and energy insecurity. Besides the humanitarian consequences, disasters such as floods also heighten investment risks due to creation of "stranded assets" such as wastewater treatment plants in developing countries. Global climate models can be used to identify the geographical distribution of disaster risk but without being able to specify the regional intensity, frequency and duration of events. In sub-Saharan Africa, the difference between what models forecast and the reality of dryness has come to be known as the "East Africa Climate paradox". Because of model inconsistency, it is difficult for investors and decision-makers to be forewarned about impending events and to respond when they occur.

Two recent policy trends have, however, converged to enhance the potential for cyber-enabled effective disaster response: (1) the expansion of regional early warning systems (REWS) and (2) a growing demand for open data platforms to enhance accountability of decision-making processes. For instance, the Africa Data Consensus (ADC) resolved to emphasize the role of open data networks to support coherent decision-making through better organization of data and models. From a scientific standpoint, agent-based modelling has also begun to advocate more forcefully for engagement of stakeholders to develop and calibrate models. In this regard, our prior research to pilot-test the Wastewater Reuse Effectiveness Index (WREI) for Sustainable Development Goal 6.3 revealed the potential of down-scaled models to advance circular economy pathways, such as wastewater recycling with potential to mitigate disaster risk.

There are three challenges that need to be overcome for disaster resilience to be enhanced. First, data in real time need to be captured from water infrastructure points, for which advances in Internet of things (IoT) such as sensors, mobile bar codes, artificial intelligence and/or remote sensing drones can be useful. This would enable data to be captured seasonally from water infrastructure points that offer a range of public services including livestock feeding points, water supply and wastewater treatment. Second, data, once captured, need to be integrated using data fusion techniques and transformed into a form that can be used by decision-makers (Mannschatz et al., 2010). Knowledge translation tools such as scenario analysis, composite indices and benchmarking of infrastructure performance can benefit from application of data fusion techniques. Finally, data infrastructure that enables regular monitoring of public service delivery and its effects on disaster resilience needs to be piloted through interventions such as labelling of water infrastructure points, procedures for updating and identification of norms for escalation of complaints within public bureaucracies in response to feedback received from consumers.

The overall aim of the Theory of Change Observatory project will therefore be to enhance regional capacity to develop, pilot-test and validate regional climate models that enable the prediction, assessment and response to effects of droughts and flood risk by

- developing a place-based observatory based on principles of dispersed data handling and reuse (Work Package [WP: 1,4]);
- co-curating and co-designing regional research that focusses on downscaling and coupling robust models of disaster risk monitoring (WP: 2,3);
- pilot-testing and validating composite indices as a means of knowledge translation with the objective of building a theory of change on disaster resilience.

Disaster resilience: horizon scanning (research design principle 1)

The United Nations Global Assessment Report on Disaster Risk (UNISDR, 2017) warns that given population growth, rapid urbanization in hazard-exposed countries and investments that do not seriously take disaster risk into account, potential future losses can be enormous. These losses may be evident through their effects on infrastructure. For example, prolonged drought may result in 'sunk costs' in water infrastructure being underutilized. In this connection, the UNISDR emphasizes the five priorities of the Hyogo Framework for Action (HFA) and the Sendai Framework which cover (1) governance- organizational, legal and policy frameworks, (2) risk identification, assessment, monitoring and early warning, (3) knowledge management and education, (4) reducing underlying risk factors and (5) preparedness for effective response and recovery.

To better appreciate the concept of disaster resilience, it is useful to examine trends in sub-Saharan Africa and South America. For example, the regional state of Tigray, Ethiopia, has been continuously suffering from drought-related disasters. The regional government has taken steps to mitigate drought and has identified drought-resilient crops in cooperation with research organizations and NGOs. However, prediction of drought is still a hurdle for planning and management purposes. This is usually associated with lack of historical meteorological, hydrological and agricultural data using the state of the art in forecasting climate variabilities and changes based on outputs of global climate models (GCMs). Therefore, research is needed that will fill the data gap through using remote sensing for which global models would have to be downscaled. Such research can have direct impact on planning and management of the natural resources and help cope with drought in the regional state (Huelsmann and Ardakanian, 2013). On the other hand, floods are a major natural disaster in South America, and in Brazil, the effect is reflected in the frequency and severity of health and socioeconomic impacts. The accelerated process of urbanization in recent decades and marked social inequality amplify the consequences and complexity of disaster risk management in the region. In recent years, efforts to increase the response capacity of early warning systems to avoid and reduce the number of disasters and losses have led to a large investment in a monitoring network and in calibrating climate models to make them fit for purpose by incorporating insights offered by regional scenarios and models. However, progress is yet to be made in building indicators and tools that can attract investments by governments, organizations, corporations, development banks and nonprofit organizations to mitigate risk and/or implement adaptation strategies, especially in large urban areas, and improve access to safe drinking water and sustainable management of environmental resources in Latin America. Therefore, collaborative research with partners from several other regions with diverse experience in disaster risk management will favor the construction of indicators and the sharing of information to improve disaster resilience in South America.

Drought resilience: concept, typologies and indices (research design principle 3)

Drought can be defined as a 'deficiency of precipitation over an extended period of time, usually a season or more which results in a water shortage for some activity, group or environmental sectors' (Gan et al., 2013). It is a temporary phenomenon and should not be confused with aridity, which is a permanent feature of climate. Drought also differs from natural hazards such as floods, earthquakes and landslides in several ways. For example, it is difficult to determine the onset and end of an event, and its duration can have a larger range extending from months to years. Another distinctive feature of drought is that its impacts are diffuse and spread slowly over a larger geographical area making them difficult to quantify because they accumulate over time. To monitor the frequency, duration and incidence of disaster risk, it is necessary to consider the level of exposure to the natural hazard and the degree of vulnerability of a society to the event. One may examine exposure to natural hazard by considering the predisposition of an area to the event and the proportion of people or assets affected by it.

Vulnerability to disasters depends on the characteristics and circumstances of a community, system or asset that makes it susceptible to the damaging effects of the event. A more comprehensive understanding of vulnerability can help regions to anticipate and improve early warning systems, help prevent disasters in future and reduce risks by enhancing resilience. We understand disaster resilience as the enhanced ability of a system, community or society that is exposed to hazards to resist, absorb, accommodate and recover from the effects of a hazard in a timely and effective manner, including through the preservation and restoration of its essential and basic functions and structures. The ability of governments to provide timely relief, invest in reconstruction and buffer economic downturns heavily influences a country's resilience to risks posed by disasters. Our research on the subject reveals that administrative decentralization, capacity of local governments and networks for information sharing among communities can prove to be critical aspects of a framework that supports enhanced disaster resilience (see Kurian et al., 2016).

Agent-based modelling and applications of the nexus framework for droughts (research design principle 4)

Experiments repeatedly find that communication bolsters cooperation but do not explain why (Poteete et al., 2010). Here, agent-based models guided by backcasting principles are key to predicting the behaviour of agents within a complex and changing political economy. The nexus approach, by offering a framework for integrative modelling of trade-offs in environmental management with the objective of advancing synergies in decision-making, offers an opportunity to study the political economy of public decision-making (Kurian et al., 2019). A prerequisite for the development, calibration and validation of coupled models of disaster resilience is the documentation of protocols and standards in agent-based modelling that would enable scholars to check and build upon each other's work. The construction of longitudinal data sets that such a process will enable can generate hypothesis to test the relationship between a development intervention (i.e., drought financing) on resilience of households and communities in the face of future events. Longitudinal research is more likely to uncover the design of light touch regulatory instruments that enhance resilience at a lower cost by assessing the efficacy of low-cost technical options as part of a larger strategy of piloting policy instruments.

Open Access Platforms (or place-based observatories) by supporting the development of such protocols and standards could enable the up scaling of research results for use by decision-makers. This perspective is radically different from a focus on decision support systems (Bui, 2000). A weakness of decision support systems is their specialized, local and thematic problem focus that does not promote cross-fertilization across regions. By contrast, place-based observatories rest on the premise that they allow for the linking of various data sources. Doing so enables the integration of data as well as closing the data gaps so long as the reliability and quality of data can be ensured. By linking already available data, it becomes possible to contribute to a 'web of data' on an issue such as drought resilience. A 'web of data' not only will allow for greater access to data and databases engenders a systematic, time- and resource-efficient analysis to identify gaps and overlaps, as well as discrepancies between various sources of the same or similar data (Shim, 2002). The potential to increase the frequency of available data also increases the potential for real-time information to match decision-making cycles associated with, for example, public sector budget cycles. Coupled with a mix of visual and data transformation tools for quantitative and qualitative research, the basis for evidence-based decision-making can be strengthened (Mannschatz et al., 2015; Kurian, 2010). Place-based observatories can potentially assist with the integration of earth observation data with other sources of data such as surveys, local registries, private, remotely sensed and big data to create multi-dimensional and downscaled models of environmental change (Terry et al., 2014).

Methodological innovations: a regional focus of benchmarking and scenario analysis (research design principle 5)

Droughts and floods are complex phenomenon, highly driven by climate and hydrologic regimes, but also affected by the interactions between the atmosphere and the land surface. This is why applications of remote sensing, artificial intelligence and virtual reality can be useful in creating anonymized data sets. The interactions between the atmosphere and the land surface are compounded by human activities, such as land use change, groundwater withdrawal and overgrazing that reduce vegetation cover, and cause desertification by lowering soil moisture content. Data from Africa shows that a reduction in soil moisture content can influence climate, possibly influencing the occurrence of drought (Gan et al., 2013). Increasing soil erosion that arises from anthropogenic pressures on environmental resources, on the other hand, can further exacerbate drought-like conditions. The environment–development nexus framework to model trade-offs can explain the interaction between biophysical-scale interactions and administrative-scale decision-making that influences human action in the form of herd sizes, crop and land use choices and adoption of conservation measures such as stall feeding of cattle or water reuse (Howarth and Monasterolo, 2016). Certain land use choices, for example, may increase the risk of soil erosion that in turn has three major effects: loss of nutrients necessary for plant growth, downstream damage to hydropower and irrigation infrastructure and depletion of water storage capacity due to sedimentation of reservoirs and streams (Rasul and Sharma, 2015; Endo et al., 2016; Yang et al., 2016). In response, what synergies involving different ministries, public sector departments, private sector and community organizations may be necessary to mitigate the risks to human populations and the environment?

Our previous research to establish the Africa Points of Excellence (APE) research consortium on drought risk monitoring highlighted the role of composite indices as a tool that can be used to visualize the intensity, frequency and distribution of risk; generate genuine stakeholder discussion on environmental, socioeconomic and institutional norms and standards and inform the design of interventions that builds disaster preparedness by visualization of hot spots- where poverty and environmental pressures coincide (Kurian et al., 2019). Robust indices include a role for remote sensed data to inform use of scenario analysis and benchmarking tools to monitor the effect of ongoing and planned interventions on levels of disaster risk (Khan et al., 2020). Our research on droughts in Africa also shows that it may be possible to take advantage of the plethora of REWS to build a Disaster Resilience Effectiveness Index that accommodates for collection and analysis of data covering biophysical, socioeconomic and institutional dimensions of disaster risk. In this regard, we found that a regional research network has much to contribute with regards to collection of data from rural water supply and livestock watering points, to create a open access platform of interoperable data fields on disaster resilience to inform decision-making that addresses the following issues (Hall and Tiropanis, 2012):

- Creation of regional maps
- Creation of multimodel ensemble-based system to account for uncertainties associated with climate predictions
- Calibration of data from REWS with data gathered from global systems to create more accurate climate predictions
- Incorporation of seasonal rainfall variability in climate predictions through integration of statistical models in frameworks that are developed based on regional data
- Incorporation of data on the atmosphere–land surface interactions in climate predictions

Research outcomes

Linked databases (research design principle 7)

Linked databases (with password access to backend and anonymized data) will focus on creating a global database on global change: demography, urbanisation and climate. It is envisioned to build links and interconnections between already available data from various data sources (open data – e.g., open source databases, reports available on the web; licensed data – e.g., password protected databases; private data – e.g., generated through a citizen observatory). Linked databases will also generate nexus analytics on different aspects of (1) water–energy food nexus, (2) water–soil–waste nexus and (3) poverty–environment nexus. Links and interconnections between already available data will contribute to a 'web of data' that is greater than the sum of its parts. Bundled into this 'web of data', it will not only allow for greater accessibility to data/databases relevant to the nexus, but also further a systematic, time- and resource- efficient way to identify gaps and overlaps, as well as possible discrepancies between different sources of the same or similar data. The Theory of Change Observatory on Drought Resilience (TOCO_DR) will explore the use of 'web crawler' software to harmonize the various data resources so that information and knowledge can be made available in a way that can further collaboration on a global scale (Lawford, 2019).

It will also contribute to aiding data visualization and evidence-based decision-making. This will be supported by a database of visualization/modelling tools, while responding to the challenges of institutional fragmentation. Linked databases will also enable the arrangement of longitudinal case studies through distributed access to portals of regional partner institutes (i.e., Makelle University, Ethiopia, University of Sao Paulo, Brazil and The Water Institute, Tanzania). Additionally, linked databases are going to address shortcomings identified during the implementation of MDGs and advance the SDGs, one goal of which is to 'increase significantly the availability of high-quality, timely and reliable data disaggregated by characteristics relevant in national contexts'. By linking data from various sources, including from different regions, cross-fertilization of ideas can occur in a more effective manner. Data sources that can provide the basis for linked databases are threefold: (1) UN agencies and university/research institutes, (2) Member States and national-level NGOs and (3) private data sets comprising data that are collected, aggregated and visualized using SMS and GIS sources. The TOCO_DR linked database function will classify data using different mediums (e.g., mobile) and from different sources (e.g., users).

Virtual learning environment (research design principles 2)

The virtual learning environment of TOCO_DR will focus on consolidating knowledge that emerges from teaching and learning activities that relate to (1) classroom or face-to-face teaching, (2) thematic online courses and (3) tailor-made training programs that respond to demands of decision-makers, practitioners and students with an interest in the planning and management of environmental resources. Integrating blended learning into the TOCO_DR permits the formation of synergies between other functionalities of the observatory. In this way, the online learning platform advances the nexus approach through transdisciplinary approaches that promote robust interfaces between education, research and policy dialogue. Towards this end, the following outcomes are envisaged:

- *Assignment archives* – e.g., case studies, which focus on the science (links between research and capacity development/training) and policy domains (dissemination of institutional practice).
- *Final portfolios* – achieved through reflections and learning objectives relating to transdisciplinary approaches that advance the nexus approach.
- *Possible research proposal identified by the course tutor* – in the form of clearly defined research questions that can become the basis for future nexus-oriented research or PhD projects.

Nexus repository (research design principle 8)

The nexus repository will comprise analyses of regional consultations and process documentation of field-testing and pilot projects of nexus planning approaches and methods (in the form of documents and videos). The TOCO_DR will analyze proceedings and presentations of various regional consultations organized by consortium partners. This analysis will generate important insights relating to needs assessments, gap analysis and overlaps in a cost-effective manner. The ambition of the TOCO_DR will be to empower scientists by providing them with a harmonized collection of new and existing data sets to be able to engage and influence the policy process. The repository of data and associated data handling protocols will foster connectivity with online learning through analysis that links course proposals and questions to thematic challenges of UN agencies and member states.

By capturing and analyzing the results of pilot projects and field-testing, approaches, methodologies and technology can be tested and evaluated. In addition, the repository of data will serve as a means for monitoring impacts of engagement with Member States and other partners, such as universities, research institutes or international organizations and NGOs as well as partnerships within the UN system. In this way, the overall aim of translating nexus research into policy, thereby informing the UN system and policy/decision-makers, can be achieved in a systematic and efficient manner. The nexus repository function will consolidate knowledge through regional consultations and piloting of new approaches to planning and environmental management.

Nexus laboratory (research design principle 6)

The nexus laboratory will employ tools classified in the linked database function to perform analysis on data sets on disaster resilience. Working papers will document the results of applications of specific tools to data sets organized for different dimensions of the nexus of water, soil and energy resources. The nexus laboratory will draw upon the analysis to identify credible data proxies for design of monitoring and evaluation frameworks for nexus pilot projects. The TOCO_DR will employ indices as a way of documenting and analyzing the changing nature of *trade-offs* in decision-making that can potentially have implications for advancing integrated disaster resilience planning and management. Indices have the potential to guide consultations at regional policy forums and dialogues at international conferences. The nexus laboratory will advance the knowledge translation and information transfer functions by applying scenario analysis and data visualization tools organized in the TOCO_DR to support benchmarking exercises grounded in comparative analysis. With the help of the TOCO_DR, implementation processes of innovative knowledge-based approaches for disaster monitoring, early warning and response can be tracked and evaluated. It also monitors progress of collaborative research projects to identify opportunities where new methodologies and policy advocacy can strengthen feedback loops at multiple levels of governance within Member States. The various TOCO_DR databases and indices, thus, offer a more dynamic means of supporting decision-making.

Network effects in the global south (research design principle 8)

Although the global south has a plethora of early warning systems, rarely are such systems properly integrated in decision-making structures of government. This is important to fully integrate disaster risk management in planning processes that inform allocation of financial and human resources. Integrated planning would necessitate sharing data between government departments, coordinated decision-making involving multiple ministries and local government jurisdictions. This potentially could mean a loss in power and authority for some entities. Yet, if synergies are forged between different stakeholders, there are network effects to be derived for the region. Here are some examples of how enhanced regional capacity for disaster risk monitoring can foster network effects in the global south:

- Regional soil and remote sensing labs can facilitate analysis for decision-makers in several provinces and countries in the region
- PhD research projects can draw upon data from ministries and labs across countries to develop regional standards of water quality and calibrate regional climate models
- Curriculum on governance aspects of rural water supply and disaster risk monitoring methods can be offered online for PhD course participants in the region
- Training of Trainer workshops for laboratory technicians and programmers of remote sensing labs can be organized by adopting a regional focus to create a standardized set of protocols that can be quickly replicated at low cost
- Regional research calls can facilitate a prioritization of ideas for which further research is required to regionalize models and calibrate methods to ensure they fit to local requirements

- International and regional conferences at Stockholm Water Week, Arab Water Week and IIT Water Conclave can serve as a platform to engage decision-makers and donors and prevent a duplication of effort in disaster preparedness.

A consortium approach: facilitating political engagement for robust monitoring (research design principle 8)

To achieve seamless interaction between science and policy, we have previously adopted a consortium approach to co-curating and co-designing research on drought monitoring in sub-Saharan Africa. A consortium approach fosters transdisciplinarity that supports integration and enhances synergies, providing the necessary evidence base for effective decision-making. To facilitate the necessary political buy-in, efficient governance structures and enable policy-relevant research, we therefore advocate for an approach that builds regional partnership involving governments, civil society and knowledge institutes. Our previous research on drought monitoring highlighted the following lessons that we will incorporate in the design of the TOCO_DR project:

Proposal writing workshop

Engages researchers in identifying nexus priorities through basic examination of relevant issues and themes (e.g., effects of drought on water and soil resources). The findings are shared with researchers within as well as across regions, enabling cross-fertilization, classification and consolidation of data and knowledge. The outcomes of a proposal writing workshop will form the basis of concept papers, which consider national conditions, demands and priorities. Bringing research in line with national, regional or international contexts (e.g., identifying the relevant disciplines, scale, boundary conditions) guarantees policy relevance, increases impact and promotes political buy-in as well as regional ownership. Here, the utilization of transdisciplinary approaches and methods is key for consolidating and translating scientific knowledge into useful information and evidence that benefits decision-makers in the development of strategies, policies as well as environmental resources management planning and implementation frameworks.

Regional consultations

Involve scientists, researchers from various disciplinary backgrounds (chosen according to the needed expertise and knowledge) as well as government officials. Engagement and collaboration of government officials is crucial at this stage. It will ease the process of translating scientific findings into policy-relevant guidance, since it allows for empirical, normative and local cultural and institutional considerations (e.g., policy formation processes) to be considered early in the process and before consortium formation. Regional consultations serve to adapt the concept papers developed during proposal writing workshops (step 1) to the respective policy context, allowing for additional comparisons and identification of challenges within a region. Determining and analysing some of the commonalities and differences between various national/local contexts will be of great advantage for effective collaboration, dialogue and monitoring of activities advancing a nexus approach to the management of water, soil and energy. In this way, concept papers become technical proposals, which may trigger the formation of a regional consortium.

Consortia that foster citizen science

Allow for mutual, transboundary learning and cross-fertilization as well as implementation of technical proposals. The formation of a consortium is based on a cooperation agreement. It constitutes the first step towards implementing the nexus approach according to national, regional and/or international demands. This is done through the transfer and translation of nexus knowledge and methodologies to support dynamic and evidence-based decision-making. Depending on the exact modalities and choices of methods, activities within a consortium may advance some or all the goals of the TOCO_DR, namely cross-fertilization, piloting and field-testing, capacity development, dissemination of insights and policy advocacy, impact monitoring and evaluation. Working with partners will not only boost institutional and individual capacities and knowledge but also enhance knowledge exchange and consolidation as well as cross-boundary and cross-regional dialogue/understanding. A consortium engages not only policy-makers but also scientists, practitioners and, if applicable, other stakeholders, who provide the research, practical experience and expertise needed for its activation, while also laying the foundation for robust citizen science that can support disaster resilience planning and management. This can potentially overcome the challenge of attribution in global public goods research that we alluded to in Chapter 2.

4. A room with two (blinkered) views

The Belmont project offers an opportunity to break out of the straitjacket imposed by conventional modelling approaches. The application of open data principles, a core feature of Belmont projects, opens up the possibility to model complex interactions between biophysical resources and public services using appropriate time and space parameters. The open data approach also allows for the proliferation of big data analytics, visualization tools and open source methods (such as R) for aggregation and synthesis of data from public and private sources, perhaps even in real time. Microsoft SQL software is one example of the computing software required to undertake such an enterprise.

But boundary organizations (global public goods research institutes and think tanks) are, however, hamstrung by the limitations imposed by their recruitment processes. Research programmers with expertise in computer science backed up by their understanding of the research and policy process are a breed of professionals in short supply today. Yet, research programmers would typify the data science professionals with boundary spanning skills that global public goods research so desperately needs. This is because conventional policy think-tanks tend to focus on policy analysts or research professionals with a blinkered view of the

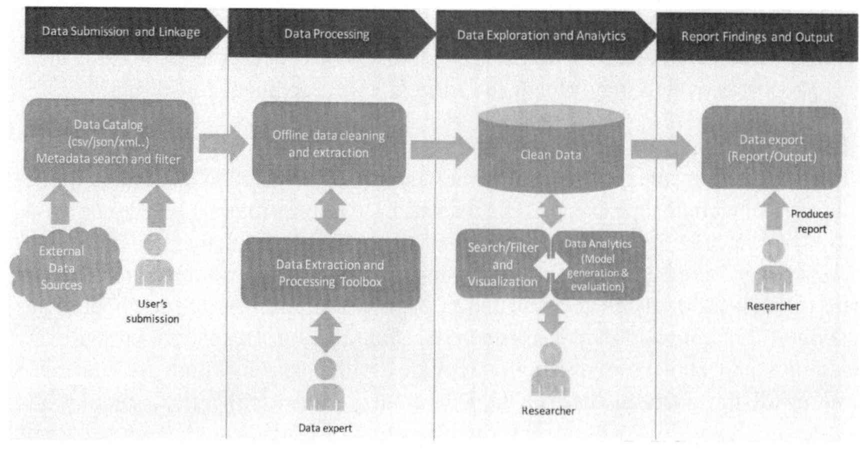

FIGURE 4.2

Computing workflow for data analytics.

complexities imposed by the environment–development nexus. Fig. 4.3 is an example of how subject matter experts tend to view a research problem as it pertains to water shortage in Loess plateau, China. They tend to assume that the answers to their research question would automatically appeal to decision-makers at the end of the research project. On the other hand, computer programmers tend to focus less on the development challenge and issues of research framing and more on issues of data submission and linkage, computing power to enable data processing, widgets that support data exploration and analytics and procedures for producing research reports (Fig. 4.2). A room with two (blinkered) views characterizes the challenge of erecting open-access platforms that incubate research questions that address critical public policy concerns. The extent to which this challenge is addressed will also determine the ability of think tanks to seamlessly experiment with backcasting approaches in environmental planning and management, an issue we turn to in the next section.

5. Environmental back-casting: re-theorizing the environment-development Nexus

The Belmont proposal that we described earlier seeks to develop a model of models on how we can learn about institutional success and failure by unleashing the power of longitudinal analysis through engagement with multiple sites via the mechanism of an Open Access Platform that provides opportunities for co-design of research, co-curation of data and online learning. The project model is focussed on droughts, but what if we expanded our scope to include other challenges such as floods? Would the model change much in terms of its focus on longitudinal analysis, multiple sites of engagement and advancement of opportunities for co-design of research, co-

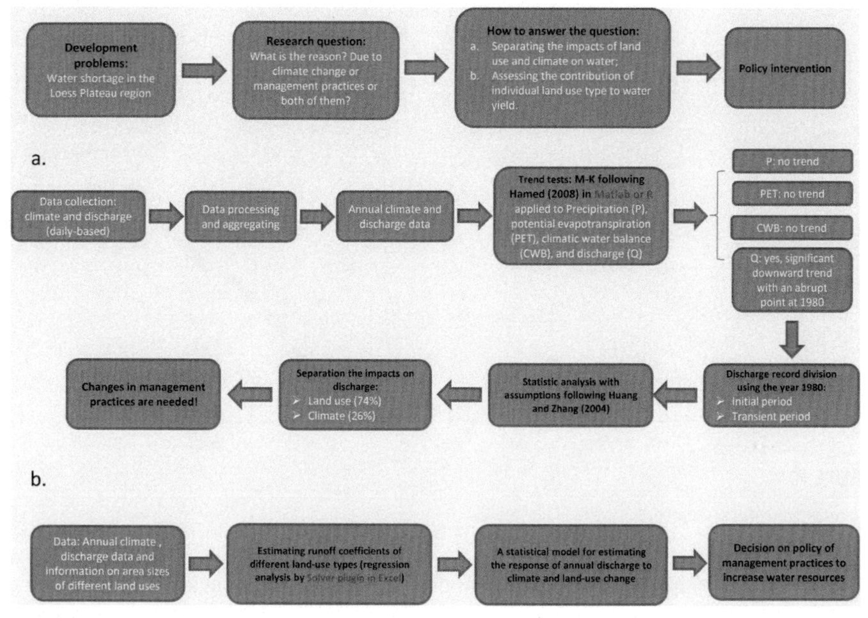

FIGURE 4.3

Scientists view of the science-policy interface.

curation of data and online learning? Are there any comparative merits that we are not aware of employing foresight and/or back-casting analyses for assessments via the mechanism of Open Access Platform (Wihlborg et al., 2019)?

The answer to the first question is no: the model of models for co-design of research, co-curation of data and online learning would not be vastly different. This is because the emphasis is on improving the research experience – setting up the interfaces that would make it possible for different data sets to be standardized, linking global models that already exist, addressing the ethics issues that arise from data sharing, archiving protocols for data sharing and facilitating online learning and optimizing upon the time and resources involved in undertaking repetitive research tasks (Fig. 4.4). There are several guiding hypotheses for such an approach to open standards research collaboration and open source modelling that could include the following: (1) more eyes on data allow for potentially more interesting questions, (2) different disciplinary perspectives allow for innovations in methods, (3) a unifying framework allows for rigorous results, (4) engagement with organizations at different scales and regions allows for more entry points into policy debates and (5) multiple sources of funding enhances applications of research to address priorities in public, private and community domains. The lessons that we learned from conceptualizing and launching the Nexus Observatory at United Nations University, Germany, based on cooperation with the GIZ further emphasized the potential of place-based observatories in undertaking assessments

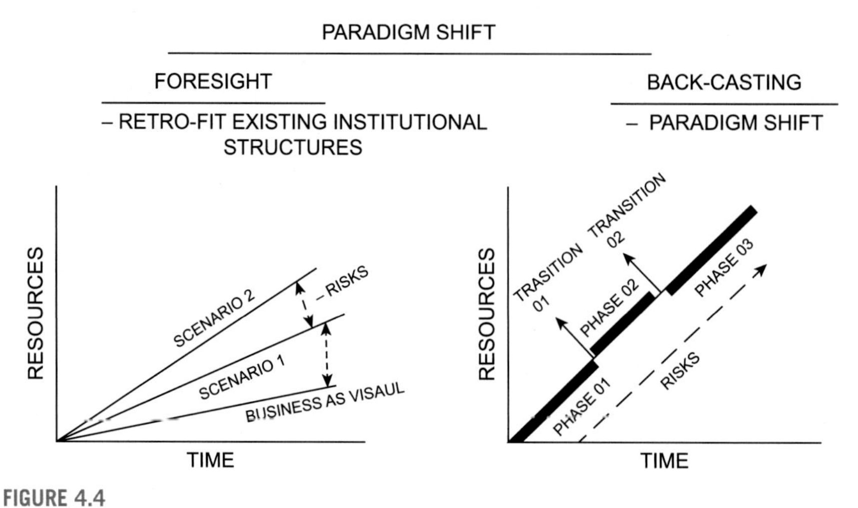

FIGURE 4.4

Paradigm shift.

of institutional trajectories in the environment–development nexus. This is supported by our experience with data integration, data analysis, data visualization and data valorization that we applied to our effort to develop and pilot-test a composite index to monitor achievement of the Sustainable Development Goal (SDG) target 6.3 (Kurian et al., 2019).

Foresight analysis or back-casting?

The Nexus Observatory project also helps us reflect upon the second question that we raised relating the comparative merits of foresight analysis versus back-casting in assessments of institutional trajectories in the environment–development nexus. Nexus analysis helps us realize the gaps that exist between normalized and imperfect data sets, planetary versus and administrative-scale perspectives of environmental resource use and management of project and budget priorities such as maximizing expenditure as against the search for the most cost-effective ways of achieving developmental goals. It is against this background that we must re-examine the concept of foresight analysis so as to promote new ways of thinking about the future (Fig. 4.4). Focussed on the analysis of environmental risks and promotion of database creation, foresight analysis by its nature would take out novelty in scientific analysis since it would make us think of a future that fits into a world that is not vastly different from the present in terms of institutional environment, biophysical resources, institutional arrangements and infrastructure. In principle, this would fit well with perspectives of incrementalism in public administration that does not allow for paradigmatic change. But will foresight perspectives be viable given the deep uncertainty and complexity that awaits us in the future (Sharmina et al., 2019)?

While the corona virus has shaken the fundamental ways in which we conventionally engage with economy, run public services and relate to each other, it may be fair to foresee similar triggers that would vastly change our future socio-economic and political landscapes with discernible impact upon institutions, infrastructure and resources. Why not then instead envision a future we would like and work backwards to decide on the specificities of institutional arrangements and infrastructure options we would like to prioritize to pursue sustainable development? This is exactly what back-casting would focus on doing and should that not be an approach, we should embrace for the study of the environment—development nexus? The focus of such an approach would be on using data to mediate dialogue among the sciences to advance a balance between modeling and the imperatives of monitoring institutional trajectories.

As you may recall, we previously highlighted four aspects that deserve attention in typology construction: institutional environment, condition of biophysical resources, institutional arrangements for delivery of public services and the frequency, intensity and duration of risks that arise from management of infrastructure. By embracing the idea of back-casting we will be able to further the cause of typologies of trade-offs that serves as the foundation for the institutional trajectory analysis. Backcasting features scenario analysis (not environmental risk analysis alone) and OAP which goes way beyond creation of simple databases. Furthermore, the approach would go a long way in emphasizing the data valorization challenge encompassing validation of data and models and protocols for data transformation. This is where the full power of artificial intelligence, virtual reality and remote sensing can be exploited to help build dynamic and multi-dimensional models of environmental resilience (Coogan, 2020). This would allow us to think of policy goals not as problems to be solved once and for all, but as norms and standards that are maintained and modified over time (Gregory, 1997: 188). Furthermore, we will have to think of modelling as an exercise that if framed well could impact upon this process of setting, modifying and monitoring policy goals.

Environmental back-casting and implications for the study of environmental policy and governance

Every policy instrument is based on a prediction — for example, a bet that travel bans will limit the spread of COVID-19. In that sense, every policy maker is a forecaster, but forecasting is made complicated by a poor understanding of the non-monotone, recursive and non-linear effects of institutions (rules of the game), incomplete information and discretionary behaviour of agents within public bureaucracies. The scope for discretionary behaviour of agents is compounded by cognitive biases which scientists normally hold about data, decision making and human behaviour. Two examples of bias are in order here. First, while trade-offs are a reminder that we cannot have it all, human beings have a natural tendency to avoid having to make difficult decisions. For example, decision-makers may eliminate the option of imposing a carbon tax[3] to combat climate change until the time has been reached when the option of a carbon tax is no longer viable since other more extreme trade-offs may have emerged that are tied to climate change (Weber, 2020). Second, studies have shown that people do not always learn and revise their beliefs based on new information and evidence. In reality, however, it is not that simple. In fact, people give more consideration to new information and evidence if that is tied to a concrete personal experience (like the death of a friend from COVID-19) than a news report about high infections rates when deciding on whether to alter their behaviour by wearing a mask and practicing social distancing.

The cognitive biases we hold are only one set of factors that prevents decision-makers from making optimal decisions. The other factor is the belief that forecasters hold that it is possible to calculate the odds of possible outcomes and thereby transform uncertainty into quantifiable risk (Scoblic and Tetlock, 2020). Nevertheless, the limits of imagination can create blind spots that policy-makers tend to fill with past experience. 'They often assume that tomorrow's dangers will look like yesterday's, retaining the same mental maps even as the territory around them has changed dramatically' (ibid:10). Without access to a real-time map of prey in different hunting grounds, for example, a prehistoric hunter might resort to a simple rule of thumb: look for animals where his fellow tribesmen found them yesterday. Similarly, forecasters would rely on calculation which is driven by data and the assumption that the future will in some way reflect the past. Another, shortcoming of foresight analysis is that it tends to work with quantifiable risk to develop compensation programs for affected communities. Such an approach smacks of a humanitarian and knee jerk response especially in low-income contexts where alternative livelihood options are limited. Instead a creative approach to designing population resettlement programs could be one example of a climate adaptation strategy that relies upon multi-dimensional and downscaled models of environmental change to design effective migration policies, for instance (see Fig. 4.5).

Environmental back-casting takes one of the imagined futures as a given and asks what conditions produced it. The most recognizable form of back-casting is a two-way matrix in which planners identify two critical uncertainties and, taking the extreme values of each, construct four possible future worlds (Scoblic and Tetlock, 2020: 13). In contrast to forecasts, back-casting would not focus on being predictive but instead emphasize the importance of posing the most relevant question, challenging conventional assumptions, shaking up mental models of how the world works and encouraging cognitive flexibility to consider outliers in statistical analysis. This approach can thrive in situations supported by open access platforms and open source modelling. Forecasters by contrast have a tendency to model phenomenon for which data are already available (via databases), thereby introducing

[3] See Gates, 2021. How to avoid a climate disaster: the solutions we have and the breakthroughs we need, Alfred Knoph.

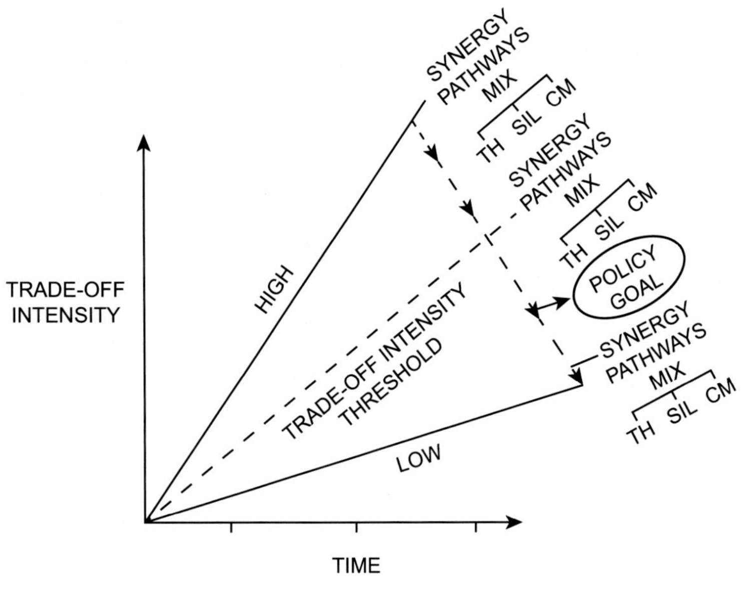

FIGURE 4.5

Backcasting synergistic pathways to public action. *CM*- critical mass; *SIL*- siloes; *TH*-thresholds.

bias for biophysical resources and by being overly focussed on certain regions of the world where English is the predominant language of science (Economist, 2020).

From a policy perspective, the greatest challenge confronting forecasters is deciding upon the most appropriate question that can inform the focus of modelling exercises. Scoblic and Tetlock (2020) point out that making a decision based on one specific forecast would be a mistake: the estimated probability of an event is a poor proxy for the significance of the event. As an example, they point out that the question 'will Putin relinquish power within the next two years?' is less consequential than a question that asks 'what would Putin's abdication of power mean for US – Russia relations?' Modelers focussed on posing the most pertinent questions enhance their chances of being able to contribute towards policy-making. By measuring the difference between estimates and the actual occurrence of events, modelers can calculate a score to show how close their model calibrations were to reality. Meteorology is an example of field where modelers have been able to exploit the advances in machine learning and artificial intelligence to link models and combine in situ data collection with big data to enable ever more precise weather forecasts that can support environmental planning and management.

A key lesson emerging from these analyses is that widespread adoption of back-casting in environmental planning and management will hinge upon identifying individuals who are naturally numerate and inquisitive, train them to think probabilistically and then group them to leverage from the wisdom of expert groups. The best back-casters would approach seemingly intractable questions by decomposing them into part, researching the past frequency of similar (if not analogous) events, adjusting the odds based on the uniqueness of the situation and continually updating their estimates as new information emerges. There are several ways in which back-casting approaches in environmental planning can advance integrative modelling on the environment–development nexus. First, instead of evaluating the likelihood of a single long-term scenario, question clusters allow researchers to break down potential futures into a series of clear milestones or phases (see Fig. 4.4 shown earlier). Preliminary answers to specific questions can provide a simple metric for judging in advance how the future is most likely to unfold – a metric that modelers can revise depending on whether the event in question materializes or not.

By contrast, foresight analysis is more knee jerk – responding with a new assessment every time a new scenario emerges in response to a new 'policy goal'. Such an approach is more risk averse because it imposes higher costs on the system every time a new policy goal is announced. In keeping with incrementalist logic, forecasting is focussed on first-generation issues of infrastructure construction and production of resources such as water and energy and is characterized by poor feedback loops with environmental and social outcomes. Back-casting-oriented research, on the other hand, is focussed on lowering the costs of monitoring environmental and social outcomes through a focus on mitigating trade-offs, feedback and social learning and is more attuned to second-order service delivery challenges, for example of resource reuse. Such an approach based on co-curation of data sourced from multiple sites is more amenable to support the piloting policy instruments – for example, a directive relating to an allocation norm for water supply for a population or an equity norm targeting wastewater or irrigation services for a population with a large proportion of poor residents. Similarly, a back-casting approach is more suited to support assessments demonstrating the applications of CCTs in enhancing uptake of technical options such as those we discussed in Chapter 3 (Agrawal, 2020).

In this connection, it is important to emphasize the usefulness of a comparative framework (*based on locally defined indicators of quality, affordability, coverage and/or service reliability*) as opposed to an evaluation framework that is guided by transcendental policy goals (for example, use of a norm such as a dollar a day for poverty alleviation or norm of 10% contribution by community towards capital costs of water projects (Sen, 1999). This is because the returns from investing in research that targets normative change may be enhanced by a series of smaller successes with regards to reform of institutional instruments and practices. Real-time information employing a comparative framework besides illuminating incremental reform processes within the public sector could also highlight processes involving citizens outside the reach of public programmes: what are their coping strategies in relation to other groups? What options exist for governments to intervene in the interest of providing affordable and reliable services to such groups? For findings of these inquires to influence policy-making, such as approach to information gathering and analysis must be multidirectional in nature (involving information flows between citizens and governments and vice versa) and is predicated upon autonomy and accountability of decision-making processes (Lane, 2000; Lipsky, 1989). This is where place-based observatories supported by an open standards framework can be crucial for conceptualizing relevant questions, optimizing upon repetitive tasks in environmental research and promoting the use of data valorization and visualization tools based on the extensive use of online learning platforms (Kurian et al., 2019).

6. Conclusions

The final chapter of this volume reflects upon the discussion in the previous three chapters to outline a proposal for establishment of open- access platforms to support environmental back-casting as a precursor to advancing a theory of change. First we have provided a rationale for promoting OAP as an effective operational tool for its capacity to capture the recursive effects of institutional feedbacks in environmental governance and its ability to address complex institutional features related to equity and inclusion in environmental decision making. Drawing upon the issue of citizen science as the foundational principle of OAP, the discussion unpacked four key components of OAP — horizon scanning, the use of open standards in research, the importance of data valorization and the need to optimize upon repetitive tasks in the research process to justify the need for investing in longitudinal analysis of the environment—development nexus at multiple sites. The example of the Belmont Forum project proposal on cyber-enabled disaster resilience was used to demonstrate how the translation of the eight research design principles can be operationalized to create political buy-in for the outputs of research. This chapter concludes by laying out the guiding hypotheses for a renewed theory of change based on environmental back-casting.

Arguing that the future will be vastly different, this chapter emphasizes the merits of environmental back-casting in informing planning processes by embracing the applications of artificial intelligence, remote sensing and virtual reality to be able to build multidimensional models that are informed by integrative analysis and geared towards responding to changes in policy goals. At an organizational level, this would translate into paradigmatic change in terms of how we evaluate the impacts of research in supporting assessments of institutional trajectories in the environment—development nexus. In addition to conventional metrics of scientific excellence, *boundary science* would necessitate attention to boundary spanning skills that were overlooked up to now. These include the capacity of individuals within organizations to promote horizon scanning, regional consultation, strategic communication, political negotiation, data valorization and knowledge translation in pursuit of the ultimate goal of boundary science which is political endorsement of results of scientific research.

In conclusion, *boundary science* promotes better synergies in environmental governance through cooperation between tiers of government and promises to support more effective responses to shifting changes in agency behaviour and partnerships with the private sector and community-based natural resources management organizations. We remain optimistic that a renewed theory of change drawing upon our three propositions on theory, methods and operation that enhances our understanding of public action on environmental governance is not out of reach. Such a theory change can promote innovations in experiential problem framing, human centered design approaches to downscaling and coupling of models of disaster resilience, data-driven learning through use of typologies and composite indices and collaborative problem solving through support for pilot testing of planning instruments in environmental management to effectively monitor the vulnerability of populations to disaster risk.

References

Agarwal, B., 1997. Environmental action, gender equity and women's participation. Dev. Change 28 (1), 1–44.

Agarwal, B., 2001. Participatory exclusions, community forestry, and gender: an analysis for South Asiaand conceptual framework. World Dev. 29 (10), 1623–1648.

Agrawal, A., 2020. Social assistance programs and climate resilience: reducing vulnerability through cash transfers. Curr. Opin. Sustain. Sci. 44, 113–123.

Baland, J., Platteau, B., 1999. The ambiguous impact of inequality on local resource management. World Dev. 27 (5), 773–788.

Brodie, J., 1994. Shifting the boundaries: gender and the politics of restructuring. In: Bakker, I. (Ed.), The Strategic Silence- Gender and Economic Policy. Zed Books-North South Institute, London.

Bui, T., 2000. Decision Support Systems for Sustainable Development: A Resource Book Applications and Methods. Springer, New York.

Coggan, P., 2020. More: A History of the World Economy from the Iron Age to the Information Age. Profile Books, London, UK.

Cornwall, A., 2003. Whose voices? whose choices? reflections on gender and participatory development. World Dev. 31 (8), 1325–1342.

Cumming, G., Epstein, G., Anderies, J., Apetrei, C., Baggio, J., Bodin, O., Chawla, S., Clemens, H., Cox, M., Egli, L., Gurney, G., Lubell, M., Magliocca, N., Morrison, T., Muller, B., Seppelt, R., Schluter, M., Unnikrishnan, H., Villamayor-Tomas, S., Weible, C., 2020. Advancing understanding of natural resource governance using socio-ecological systems framework: a post-ostrom research agenda. Curr. Opin. Sustain. Sci. 44, 26–34.

Cvitanovic, C., Lof, M., Norstrom, A., Reed, M., 2018. Building university-based boundary organizations that facilitate impacts on environmental policy and practice. Plos One 13 (9), e0203752.

Economist, 2020. Starving for knowledge-big economies get the most research, but economists prioritize countries with abundant data and ignore petro-states, p. 89. December 12.

Eftelioglu, Z.J., Tang, X., Shekhar, S., 2017. The Nexus of food, energy and water resources: visions and challenges in spatial computing. In: Griffith, D.A., et al. (Eds.), Advances in Geocomputation. Springer International Publishing, Switzerland.

Endo, A., Burnett, K., Orencio, P., Kumazawa, T., Wada, C., Ishii, A., Tsurita, I., Taniguchi, M., 2016. Methods of the water-energy-food nexus. Water 7.

Gan, T., Ito, M., Huelsmann, S., 2013. Drought, Climate and Hydrological Conditions in Africa: An Assessment on the Applications of Remotely Sensed Geospatial Data and Various Models. UNU-FLORES, Dresden. Working Paper No. 01.

Gates, B., 2021. All chapters. How to Avoid a Climate Disaster: The Solutions We have and the Breakthroughs We Need, 1st. Alfred Knoph, New York, pp. 1−256.

Gevelt, T., 2020. The water-energy-food nexus- bridging the science-policy divide. Curr. Opin. Environ. Sci. Health 13, 6−10.

Gregory, R., 1997. Political rationality or incrementalism? In: Hill, M. (Ed.), The Policy Process-A Reader. Prentice Hill, Essex, pp. 175−191.

Hall, W., Tiropanis, T., 2012. Web-evolution and web science. Comput. Network. 56, 3859−3865.

Harding, S. (Ed.), 1987. Feminism and Methodology. Indiana University Press, Bloomington.

Harwood, S., 2018. In search of a WEF nexus approach. Environ. Sci. Pol. 83, 79−85. https://doi.org/10.1016/j.envsci.2018.01.020.

Howarth, C., Monasterolo, I., 2016. Understanding barriers to decision making in the UK energy −food- water nexus- the added value of interdisciplinary approaches. Environ. Sci. Pol. 61, 53−60.

Huelsmann, S., Ardakanian, R., May 6−8, 2013. Proceedings of the Regional Workshop on Establishment of a Network for Partnership of UNU-FLORES Based in Maputo, Mozambique.

Jackson, C., 1993. Doing what comes naturally? women and environment in development. World Dev. 21 (123), 1947−1963.

Kabeer, N. (Ed.), 2005. Inclusive Citizenship: Meanings and Expression. Zubaan Publisher.

Keller, E.F., Grontkowski, C., 1996. The mind's eye. In: Keller, E.F., Longino, H.E. (Eds.), Feminism and Science. Oxford University Press, Oxford/New York.

Khan, Z., Wild, T., Vernon, C., Miller, A., Hejazi, M., Clarke, L., MirallesWilhelm, F., Castillo, R., Moreda, F., Bereslawski, J., et al., 2020. Metis-a tool to harmonize and analyze multi-sectoral data and linkages at variable spatial scales. J. Open Res. Software 8, 10. https://doi.org/10.5334/jors.292.

Kurian, M., 2010. Making sense of human-environment interactions. In: Kurian, M., McCarney, P. (Eds.), Peri-urban water and sanitation services − policy, planning and method. Springer, Dordrecht.

Kurian, M., Portney, K., Rappold, G., Hannibal, B., Gebrechorkos, S., 2018. Governance of Water-Energy-Food Nexus: A Social Network Analysis Approach. In: Huelsmann, Ardakanian (Eds.), Monitoring and Implementing the Nexus Approach to achieve Sustainable Development. UNU-Springer, Switzerland, pp. 125−147.

Kurian, M., Reddy, V., Scott, C., Alabaster, G., Nardocci, A., Portney, K., Boer, R., Hannibal, B., 2019. One Swallow does not make a Summer- Siloes, Trade-offs and Synergies in the Water-Energy-Food Nexus, Frontiers in Environmental Science, Special Issue on "Achieving Water-Energy-Food Nexus Sustainability- a Science and Data Need or a Need for Integrated Public Policy?" (Editors: Rabi Mohtar, Jillian Cox and Richard Lawford), 7 (32), 1−17.

Kurian, M., 2020. Monitoring versus modelling water-energy-food interactions: how place-based observatories can inform research for sustainable development. Curr. Opin. Sustain. Sci. 44, 35−41.

Kurian, M., Veiga, L., Ardakanian, R., Meyer, K., 2016. Resources, Services and Risks- How Can Data Observatories Bridge the Science- Policy Divide in Environmental Governance? Springer Briefs, Switzerland.

Lane, J., 2000. The Public Sector: Concepts, Models and Approaches. Sage, London.
Lawford, R., 2019. A design for a data and information service to address the knowledge needs of the water-energy-food (W_E_F) Nexus and strategies to facilitate its implementation. Front. Environ. Sci. 7, 56.
Lipsky, M., 1989. Street Level Bureaucracy, in the Policy Process- A Reader. In: Hill, M. (Ed.). Prentice Hill, Essex, pp. 175–191 (London, Sage).
Mannschatz, T., Buchroithner, M., Huelsmann, S., 2015. Visualization of water services in Africa: data applications for nexus governance. In: Kurian, M., Ardakanian, R. (Eds.), Governing the Nexus- Water, Soil and Waste Resources Considering Global Change. Springer, Dordrecht.
Mies, M., Shiva, V., 1993. Ecofeminism. Fernwood Publications, Halifax, N.S.
Mohammadpur, P., Mahjabin, T., Fernandez, J., Grady, C., 2019. From national indices to regional action- an analysis of food, energy, water security in Equador, Bolivia and Peru. Environ. Sci. Pol. 101, 291–301.
Mosse, D., Farrington, J., Rew, A., 1998. Development as Process: Concepts and Methods for Working with Complexity. Routledge, London.
Muldoon-Smith, K., Greenhalgh, P., 2019. Suspect foundations: developing an understanding of climate related stranded in the global real estate sector. Energy Res. Soc. Sci. 54, 60–67.
OECD, 2008. Handbook of Constructing Composite Indicators: Methodology and User Guide. Organization for Economic Cooperation and Development, Paris.
Poteete, A., Janssen, M., Ostrom, E., 2010. Working together – collective action, the commons and multiple methods. In: Practice. Princeton University Press, Princeton, NJ.
Rasul, G., Sharma, B., 2015. The nexus approach to water- energy-food security: an option for adaptation to climate change. Clim. Pol. 16 (6), 682–702.
Razavi, S., 1997. Fitting gender into development institutions. World Dev. 25 (7), 1111–1125.
Regueiro, M., Branch, L., Derlindati, E., Gasparri, I., Marinaro, S., Nanni, S., Godoy, L., Rodrigeuz, M., Soto, J., Talamo, A., 2020. Open standards for conservation as a tool for linking research and conservation agendas in complex socio-ecological systems. Curr. Opin. Sustain. Sci. 44, 6–15.
Ribeiro, J., Da Silva, S., Neiva, S., Soares, T., Montenegro, C., Deggau, A., Amorin, W., Lopez, C., Junior, C., Salguerinho, C., Guerra, O., 2020. A proposal of a balanced score card to the water-energy-food nexus approach: Brazilian food policy in the context of the sustainable development goals. Stochastic Environ. Res. Risk Assess. https://doi.org/10.1007/s00477-020-01769-1.
Sen, A., 1999. Development as Freedom. Oxford University Press, USA/UK.
Scoblic, J., Tetlock, P., 2020. A better crystal ball- the right way to think about the future. Foreign Aff. 99 (6), 10–18.
Sharmina, M., Ghanem, D., Browne, A., Hall, S., Mylan, J., Petrova, S., Wood, R., 2019. Envisioning surprises: how social sciences could help models represent 'deep uncertainty' in future energy and water demand. Energy Res. Soc. Sci. 50, 18–28.
Shim, J.P., 2002. Past, present and future of decision support technology. Decis. Support Syst. 33, 111–126.
Shiva, V., 1988. Staying Alive: Women, Ecology and Survival in India. Kali for Women, New Delhi.
Squire, J., 1999. Gender in Political Theory. Polity Press, Cambridge UK.
Sultana, F., 2009. Commuity and participation in water resources management: gendering and naturing development debates from Bangladesh. Trans. Inst. Br. Geogr. 34, 346–363.

Terry, R.F., Salm, J.F., Nannei, C., Dye, C., 2014. Creating a global observatory for health R&D. Science 345 (6202), 1302–1304.

Tian, H., Lu, C., Pan, S., Yang, J., Miao, R., Ren, W., et al., 2018. Optimizing resource use efficiencies in the food-energy-water nexus for sustainable agriculture: from conceptual model to decision support tool. Curr. Opin. Environ. Sustain. 33, 104–113. https://doi.org/10.1016/j.cosust.2018.04.003.

Uden, D., Allen, C., Munoz Arriola, F., Ou, G., Shank, N., 2018. A framework for assessing socio-ecological trajectories and traps in intensive agricultural landscapes. Sustainability 10, 1646.

UNISDR, 2017. The Atlas- Unveiling Global Disaster Risk. United Nations Office for Disaster Risk Reduction, Geneva.

Urbinatti, A., Fontana, M., Stirling, A., Giatti, L., 2020. 'Opening up' the governance of water-energy-food nexus: towards a science-policy-society interface based on hybridity and humility. Sci. Total Environ. 744, 140950. https://doi.org/10.1016/j.scitotenv.2020.140945.

Vasconcelos, C., Schneider, S., Peppoloni, S., 2020. Teaching Geo-Ethics: Resources for Higher Education. University of Porto, Portugal.

Villamayor, S., Oberlack, C., Epstein, G., Partelow, S., Roggero, M., Kellner, E., Tschopp, M., Cox, M., 2020. Using case study data to understand SES interactions: a model-centered meta-analysis of SES framework application. Curr. Opin. Sustain. Sci. 44, 48–57.

Weber, E., 2020. Heads in the sand: why we fail to foresee and contain catastrophe. Foreign Aff. 99 (6), 20–26.

Wihlborg, M., Sorenson, J., Olsson, J., 2019. Assessment of barriers and drivers for implementation of blue-green solutions in Swedish municipalities. Environ. Sci. Pol. 203, 706–718.

Yang, E., Wi, S., Ray, P., Brown, C., Abedalrazq, K., 2016. The future nexus of the Brahmaputra river basin: climate, water, energy and food trajectories. Global Environ. Change 37, 16–30.

Epilogue

Reinstating multilateralism: scientific research and international development cooperation

The news that Pfizer had developed a vaccine with a high probability of defeating COVID-19 was welcomed worldwide with a sense of nervous jubilation in late 2020. This came against the sharp backdrop of the departure of the United States from the World Health Organization (WHO), one of the most prominent United Nations (UN) organizations at the forefront of efforts to coordinate the global response to the pandemic. The new US administration has since rejoined the WHO and returned to the Paris Climate agreement that was mediated by the UN in 2015. The Biden worldview reiterates a core principle of multilateralism – an openness to cooperate to defeat common global challenges despite our differences. Multilateralism relies crucially on global public goods research organizations (we called them boundary organizations in this book) to marshal the evidence and forge cooperation to address global challenges such as environment induced mass migration. The United Nations University (UNU), which is comprised of numerous institutes around the world, is an example of a global public goods research organization mandated to serve as the think tank for the UN and Member States that has remained remarkably silent on discussions relating to pandemic prediction and vaccine research and development for COVID-19. This should give us pause to rethink the role of global public goods research in advancing multilateral development. A closer look at conventional practices that has infiltrated public goods research organizations could explain why such entities have lost their appeal in recent years and hence their ability to cast the seeds for greater societal impact. There are persistent structural and ethical problems reinforced by a partisan world view, which goes hand in hand with political judgements that guarantee continuous support for research initiatives irrespective of their authenticity and contribution to addressing emerging policy challenges. It is high time that we engage in undoing such practices.

As the commitment to multilateral action is being renewed once more, it is an opportune time to reflect on the terms of UN engagement. Besides public health, robust models are needed to contribute towards shaping international cooperation to address another pressing problem – climate change which increasingly exhibits interconnections with other development issues such as gender violence, refugees and migration. In this connection, boundary organizations (policy oriented think tanks), which hold a particularly special place with regards to contributing towards addressing some of the world's pressing challenges, must once again reexamine what makes them different from conventional universities and establish more effective procedures in seeking out individuals with the vision, training and diplomatic skills needed to translate the noble ideals of boundary organizations into knowledge and evidence that spurs decision-makers to act decisively on the basis of robust science. Greater accountability across the rank and file of global public goods organizations is required to ensure that future funding translates into verifiable and tangible outputs such as a global model that can predict the outbreak of future pandemics, the establishment of a mechanism that supports equitable distribution of the benefits of public goods research like a vaccine that safeguards human health, improved crop varieties that can reduce hunger or an open-source platform to support transdisciplinary cooperation on the applications of artificial intelligence, remote sensing and big data analytics in enhancing regional preparedness to address environmental challenges such as droughts and floods.

While these are examples of the ideal role that global public goods research can play in advancing a sustainable and secure future, their achievement is in no small measure predicated upon member states, scientists and administrators overcoming some of their darker impulses, which often bends them towards self-interest at the cost of promoting the greater common good. Our book is developed in the hope of bending the moral arc of global public goods research towards activities that are geared to making development interventions more inclusive, cost-effective and impactful for the benefit of vulnerable and marginalized populations the world over.

Mathew Kurian, PhD
Yu Kojima, PhD
Pennsylvania
United States

Index

'*Note*: Page numbers followed by "f" indicate figures, "t" indicate tables and "b" indicate boxes.'

F

G

H

I

U

V

W

Printed in the United States
by Baker & Taylor Publisher Services